Push your Career Publish your Thesis

Science should be accessible to everybody. Share the knowledge, the ideas, and the passion about your research. Give your part of the infinite amount of scientific research possibilities a finite frame.

Publish your examination paper, diploma thesis, bachelor thesis, master thesis, dissertation, or habilitation treatises in form of a book.

A finite frame by infinite science.

Infinite Science Publishing

An Imprint of
Infinite Science GmbH
MFC 1 | Technikzentrum Lübeck
BioMedTec Wissenschaftscampus
Maria-Goeppert-Straße 1
23562 Lübeck
book@infinite-science.de
www.infinite-science.de

Editor

Thorsten M. Buzug
Institute of Medical Engineering
University of Lübeck
buzug@imt.uni-luebeck.de

Reihe: Medizinische Ingenieurwissenschaft und Biomedizintechnik

Diese Reihe umfasst Werke der Medizinischen Ingenieurwissenschaft und Biomedizintechnik, deren Themen strategisch unter den Zukunftstechnologien mit hohem Innovationspotenzial anzusiedeln sind. Als wesentliche Trends dieser Forschungsgebiete, sind die Schlüsselbereiche Computerisierung, Miniaturisierung und Molekularisierung zu nennen. Bei der Computerisierung sind dabei die inhaltlichen Schwerpunkte beispielsweise in der Bildgebung und Bildverarbeitung gegeben. Die Miniaturisierung spielt unter anderem bei intelligenten Implantaten, der minimalinvasiven Chirurgie aber auch bei der Entwicklung von neuen nanostrukturierten Materialien eine wichtige Rolle, und die Molekularisierung ist in der regenerativen Medizin aber auch im Rahmen der sogenannten molekularen Bildgebung ein entscheidender Aspekt. Forschungs- und Entwicklungspotenzial werden auch der Biophotonik und der minimal-invasiven Chirurgie unter Berücksichtigung der Robotik und Navigation zugeschrieben. Querschnittstechnologien wie die Mikrosystemtechnik, optische Technologien, Softwaresysteme und Wissenstechnologien sind dabei von hohem Interesse.

Klaas Bente

Implementation of a Magnetic Particle Imaging System for a Dynamic Field Free Line

Medical Engineering Science and
Biomedical Engineering — Volume 14

Editor: Thorsten M. Buzug

Infinite Science
Publishing

© 2016 Infinite Science Publishing
der BioMedTec Wissenschaftsverlag Lübeck

Ein Imprint der Infinite Science GmbH,
MFC 1 | BioMedTec Wissenschaftscampus
Maria-Goeppert-Straße 1
23562 Lübeck

Cover Design, Illustration: Uli Schmidts, metonym
Copy Editing: University of Lübeck, Institute of Medical Engineering

Publisher: Infinite Science GmbH, Lübeck, www.infinite-science.de
Print: BoD, Norderstedt

ISBN Paperback:978-3-945954-16-4

Bibliografische Information der Deutschen Nationalbibliothek:
Die Deutsche Nationalbibliothek verzeichnet diese Publikation in der Deutschen Nationalbibliografie; detaillierte bibliografische Daten sind im Internet über http://dnb.d-nb.de abrufbar.

Bibliographic information published by the Deutsche Nationalbibliothek
The Deutsche Nationalbibliothek lists this publication in the Deutsche Nationalbibliografie; detailed bibliographic data are available in the internet at http://dnb.d-nb.de.

Abstract

A new tomographic imaging modality called magnetic particle imaging (MPI) has been proposed in 2005. Using the non-linear magnetization curve of specific nanoparticles, a detectable signal proportional to the concentration of these particles can be generated. For medical imaging the tracer material can be injected into the blood system. Due to a fast image acquisition time, not only morphological, but also functional imaging is possible. The feasibility of medical imaging with this technique could be demonstrated at real time imaging of a beating mouse heart. Spatial encoding is provided by superimposing dedicated external magnetic fields. For the resulting shape of this field a field free line has been proposed. This work presents the implementation of the MPI system with a discretely rotatable and dynamically translatable field free line.

Zusammenfassung

Ein neues tomografisches Bildgebungsverfahren, genannt Magnetic Particle Imaging (MPI), wurde 2005 vorgestellt. Mit Hilfe der nichtlinearen Magnetisierungskurve von spezifischen Nanopartikeln kann ein Signal, proportional zur Partikelkonzentration erzeugt werden. Medizinische Bildgebung wird möglich, wenn das Tracermaterial in die Blutlaufbahn eines Patienten injiziert wird. Dank der schnellen Aufnahmezeit ist nicht nur morphologische, sondern auch funktionelle Bildgebung möglich. Die Umsetzbarkeit wurde anhand einer Echtzeitaufnahme eines schlagenden Mausherzens gezeigt. Zur räumlichen Kodierung wird die Überlagerung von externen magnetischen Feldern genutzt. Eine von zwei bisher verwendeten Varianten dieses resultierenden externen Feldes hat die Form einer feldfreien Linie. Die vorgelegte Arbeit präsentiert die Implementierung des ersten MPI systems mit einer diskret rotierbaren und dynamisch translierbaren feldfreien Linie.

Contents

Chapter 1

Introduction

Extending the scope of the human eye beyond its natural limit has been a challenge for scientists during the last millennium. In the last decades the interest of applying techniques with this purpose in the fields of medicine has been growing rapidly. The discovering of x-rays, the following development of computed tomography and the invention of ultra-sound and magnetic resonance imaging (MRI) are examples for this.

A new approach, called magnetic particle imaging, has been made in 2005 by Gleich and Weizenecker [1]. Here, the utilization of magnetic nanoparticles is proposed to generate tomographic images. With a composition of external magnetic fields and tracers with a nonlinear magnetization curve, signals can be generated and spatially encoded [2].

For clinical imaging, the particles can be injected into a patients blood system and hence morphological and functional imaging concerning this system can be achieved [3–6]. After this, they are decomposed inside the liver, where imaging might also be prospective [7, 8]. The agent's harmlessness has already been proven, since the in MRI utilized contrast agent Resovist provides the desired magnetic characteristics [9–13].

The original invention of MPI proposed a spatial encoding scheme with a so called field free point (FFP), where a particle signal is generated [8, 11]. Here a special coil assembly generates a magnetic field, where a single point remains field free and its movement causes the nanoparticles at its position to generate a characteristic signal. For a complete scan, the FFP is moved over the full field of view (FOV). An alternative encoding scheme, the concept of the field free line (FFL), was proposed and greatly improved afterwards [14–16]. By creating a coil assembly that generates a magnetic field with an FFL, a better signal to noise ration (SNR) can be achieved for a scan. To accumulate enough data to reconstruct the image, the FFL has to be rotated slowly and translated rapidly over the region of interest (ROI) [17, 18].

This has experimentally not been possible until today. Attempts have been made to create a static FFL with a moving probe and first successful reconstructions were achieved [19]. The principle practicability of a rotating FFL and a static scanner has also already been shown [20]. But for the first time, the presented scanner is able to rotate the FFL, translate it over the field of view and record a particle signal. The scanner is constructed with a gateway diameter of 30mm to provide the possibility of scanning a mouse after first test results.

Since the signal generating field and the signal itself consist of electromagnetic waves, the electronics of the scanner are fundamental to run a scanning progress. Disturbing signals have to be damped and shielded, the quality of the input signal has to be very high and the particle signal has to be filtered accurately.

The author has designed the scanner in a previous work. The field generating coil assembly has been successfully implemented in the master thesis [21]. This work presents mainly the signal generation and detection of the system. Before the computer generated signal reaches the coils, an amplification and a filtering step are implemented and the desired fields are generated. In addition to that the receiving chain, following the receiving coil has been designed and implemented. A filter is followed by an amplification unit and the signal is transmitted to the computer.

Chapter 2

Fundamentals

To present a complemented work, not only the basics of MPI are explained, but all the physical laws used in this thesis are briefly explained. These range from fundamentals in electricity and magnetism to the idea of MPI and to practical imperatives such as shielding and electric networks.

2.1 Physics

All three important parts of the imaging system MPI, signal generation, spatial encoding and signal detection are performed by electromagnetic waves. Furthermore, for the comprehension of spatial encoding by the FFL, the theory of magnetic fields is important. For signal generation, the concept of magnetism in matter plays the central role and for signal detection the law of induction has to be understood. The presented fundamentals in physics are mainly based on [22].

2.1.1 Electric charge and current

To understand the principles of electric current, it is helpful to start with the phenomenon of electric charge. The electric charge Q is a fundamental quantity of matter. Every charge consists of the sum of discrete elementary electric charges and is therefore quantized:

$$Q = \sum_{i=1}^{k} a_i e \tag{2.1}$$

with $a_i \in \{-1, 1\}$. The elementary electric charge e is derived below. The formula states that every matter has an electric charge, that can be a positive or a negative multiple of e or exactly zero. Bodies with $Q > 0$ are called positively charged, bodies with $Q < 0$ negatively charged and bodies with $Q = 0$ electrically neutral. It is notable that the sum can be zero not only, when there are no contributions to the sum, but also when the positive and negative contributions are equally frequent.

Elementary charge

To find an reference value for electric forces, a reference experiment has to be given that determines the unit in whose multiples can be calculated. The unit in electrics is the Coulomb (C), which corresponds to Ampere seconds in SI units (As). It is defined by the charge, that each of two equal electric charges with a distance of 1 m in vacuum have, when a force of

$$F = \frac{1}{4\pi\varepsilon_0}\mathrm{N} \tag{2.2}$$

occurs between them. The nature constant ε_0 is the vacuum permittivity:

$$\varepsilon_0 = 8.8543 \cdot 10^{-12}\frac{\mathrm{A^2s^2}}{\mathrm{Nm^2}} \,. \tag{2.3}$$

Based on this, the smallest and hence fundamental electric charge is

$$e = 1.602 \cdot 10^{-19}\mathrm{C} \,. \tag{2.4}$$

The carriers of these fundamental charges are called electrons and protons. Thereby electrons carry a negative and protons a positive fundamental charge.

Electric current

If the charge changes over time in a defined volume, this flow of elementary charges defines the electric current

$$I = \frac{\mathrm{d}Q}{\mathrm{d}t} \,. \tag{2.5}$$

The unit of the electric current is Ampere (A). For an alternative and more general definition, the current density $\mathbf{j}(\mathbf{r})$, that passes through an area A can be used

$$I = \int_A \mathbf{j}(\mathbf{r})\,\mathrm{d}a \,. \tag{2.6}$$

Conservation of electric charge

In a closed system, the electric charge is conserved. That means that it can neither be created nor destroyed, which is a basic requirement for Kirchhoff's circuit laws, which are introduced in section 2.3. To formulate this law mathematically, the definition of the electric charge density $\rho(\mathbf{r})$ is needed and indirectly given by

$$Q = \int_V \rho(\mathbf{r}) \, d^3\mathbf{r} \,. \tag{2.7}$$

With this, the conservation of electric charge can be formulated in the continuity equation

$$\frac{\delta \rho}{\delta t} + \mathrm{div}\mathbf{j} = 0 \,. \tag{2.8}$$

2.1.2 Coulomb's law

For a distance between two charges, high enough to assume point charges and the two carriers are not moving, Coulomb's law

$$\mathbf{F}_C = \frac{1}{4\pi\varepsilon_0\varepsilon_r} \frac{q_1 \, q_2}{|\mathbf{r}_1 - \mathbf{r}_2|^2} \, \mathbf{e}_{12} \tag{2.9}$$

can be utilized. Here $\mathbf{e}_{12}$ is the unit vector pointing from $\mathbf{r}_2$ to $\mathbf{r}_1$. This law defines the unit Coulomb and gives the relation between the electric force $\mathbf{F}_C$ and the distance of two point charges q_1 and q_2, which is inversely quadratic. To apply the law not only in vacuum, the relative permittivity ε_r specifies the force for different materials and is hence a material constant. This law can be extended to more charges by the superposition of the individual forces and the description of the charges as charge densities $\rho(\mathbf{r}_i)$:

$$\mathbf{F}_C = \frac{1}{4\pi\varepsilon_0\varepsilon_r} \int \frac{q\rho(\mathbf{r}_i)}{|\mathbf{r}_1 - \mathbf{r}_i|^2} \, \mathbf{e}_{1i} d^3\mathbf{r}_i \,. \tag{2.10}$$

Electric potential and voltage

The potential energy, resulting from a force is the line integral of that force between two points in space. Since the line integral over the Coulomb force depends only on the start and end point and not on the path of integration, the electric potential is a conservative force. Hence an electric potential energy E_{pot} can be derived

$$\mathbf{F}_C = -\Delta E_{pot} \,. \tag{2.11}$$

The electric potential Φ for a position A is defined as the potential energy per electric charge

$$\Phi_A = \frac{E_{pot}(A)}{q} \,. \tag{2.12}$$

The reason for the movement of charge carriers is a difference in this potential, since a potential difference causes a force on the charge. This difference has its own quantity, the voltage

$$U_{AB} = \Phi_A - \Phi_B. \tag{2.13}$$

The unit for the electric potential and the voltage is Volt (V; in SI: $1 \text{ V} = \frac{\text{Nm}}{\text{As}}$).

With this quantity it is easy to determine the energy, that has to be applied to move a point charge from a point B to a point A. Based on equation (2.11), it follows directly, that

$$
\begin{aligned}
W_{AB} &= -\int_B^A \mathbf{F} \, d\mathbf{r} \\[2mm]
&= q \int_B^A d\Phi \\[2mm]
&= q \left[\Phi(A) - \Phi(B) \right] \\[2mm]
&= q U_{AB}.
\end{aligned}
\tag{2.14}
$$

Since this formula only holds validity in vacuum, section 2.1.2 introduces the behavior of matter with applied electric potentials.

Electric field

Introducing the electric field $\mathbf{E}(\mathbf{r})$ is helpful for further derivations, although it is no actually measurable size. It has the same direction as the Coulomb force, but the absolute value is normalized by the only theoretically present probe charge q at a position $\mathbf{r}_1$. This means that if an electric field is known, it is easy to determine the direction and absolute value of the Coulomb force on any charge, placed inside.

A formal definition of the electric field is

$$\mathbf{E} = \lim_{q \to 0} \frac{\mathbf{F}}{q}. \tag{2.15}$$

The SI-unit of the field is $[E] = \frac{\text{N}}{\text{C}} = \frac{\text{V}}{\text{m}}$.

Starting with Coulomb's law given by equation (2.9) it can be seen that

$$\mathbf{E} = \frac{1}{4\pi\varepsilon_0\varepsilon_r} \int \frac{\rho(\mathbf{r}_i)}{|\mathbf{r}_1 - \mathbf{r}_i|^2} \, \mathbf{e}_{1i} d^3\mathbf{r}_i. \tag{2.16}$$

The vector in the integral can be written as

$$\frac{\mathbf{e}_{1i}}{|\mathbf{r}_1 - \mathbf{r}_i|^2} = \frac{|\mathbf{r}_1 - \mathbf{r}_i|}{|\mathbf{r}_1 - \mathbf{r}_i|^3} = -\nabla_{r_1} \frac{1}{|\mathbf{r}_1 - \mathbf{r}_i|} . \tag{2.17}$$

This means that the electric field is a gradient field

$$\mathbf{E}(\mathbf{r}) = -\nabla \Phi(\mathbf{r}) . \tag{2.18}$$

This gives a less phenomenological definition of the electric potential than the definition, given in 2.12. Furthermore every gradient field is free of rotation and for the electric field follows that

$$\text{rot}(q\mathbf{E}) = 0 . \tag{2.19}$$

Further for the divergence of the electric field is given by

$$\text{div}(\mathbf{E}(\mathbf{r})) = \frac{1}{\varepsilon_0} \rho(\mathbf{r}) . \tag{2.20}$$

For a detailed derivation of (2.20) see [22], pp.58.

Conductor and isolator

When Coulomb forces emerge inside matter, two cases can occur which are of relevance for this work. The material is either a conductor or an isolator. When the material is a conductor, an applied voltage and the accompanying Coulomb force cause a movement of charge carriers and according to (2.5) a current. If this movement is not generated, the material is called isolator.

For solid matter this can be roughly explained by the model of the Bohr atom. As shown in Figure 2.1, the atom nuclei are in fixed positions inside the crystal structure of the solid material. The only interchangeable particles between the atoms are the electrons, which carry a negative charge. If the single atoms each have at least one electron, whose binding energy to the atom nucleus is low enough to be exceeded by the Coulomb force, a flow of charge carriers is possible. Then the material is a conductor. A material without these electrons can not be a medium for a current and is an isolator.

Ohm's law and power loss

Conductors can be divided into two groups. The first group contains the conductors with a linear relation between applied voltage and flowing electric current

$$R = \frac{U}{I} . \tag{2.21}$$

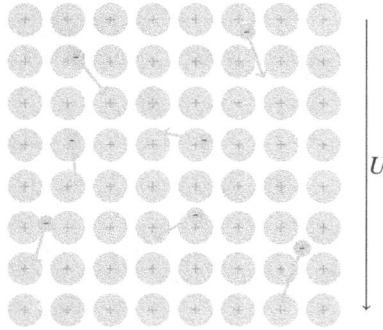

Figure 2.1: Simplified atomic model, to visualize the movement of the fundamental electric charge carriers, the electrons, inside a conductor. The favored movement direction goes along the potential difference U.

This relation is called Ohm's law and the quotient R is called the resistance and has the unit Ohm (Ω). These conductors are called ohmic conductors. Conductors in the second group feature a non-linear relation between U and I and are called non-ohmic conductors. Both cases are displayed in Figure 2.2.

In this work, the resistances of individual components of the MPI setup are of high importance. To find an optimal positioning of the assembly elements, an optimization algorithm worked with the resistances and currents inside these elements to find positioning with low heat development. The heat development and the resistances are closely related. As shown in equation (2.14), the energy needed to move an infinitesimal charge $\mathrm{d}q$ through the potential difference V is $\mathrm{d}W_e = \mathrm{d}q \cdot U$. The rate of this conversion defines the power

$$P = \frac{\mathrm{d}W_e}{\mathrm{d}t} = \frac{\mathrm{d}q \cdot U}{\mathrm{d}t} \,. \tag{2.22}$$

From equation (2.5) it is given that $\frac{\mathrm{d}q}{\mathrm{d}t} = I$ and therefore

$$P = U \cdot I. \tag{2.23}$$

With this definition of electrical power loss and with Ohm's law, it can be written that

$$P = I^2 \cdot R \,. \tag{2.24}$$

This was used in the optimization algorithm (3.1), where the coil positioning for the presented scanner was found.

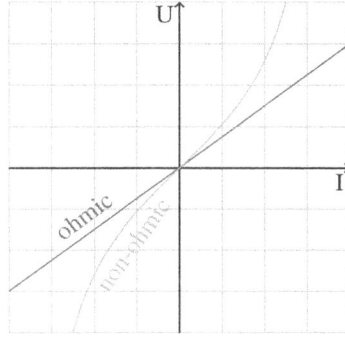

Figure 2.2: Ohmic and non-ohmic relation between a potential difference at two ends of a conductor and the current flow inside this conductor.

2.1.3 Magnetism in vacuum

The theory of magnetism is important to understand the principles of MPI. The derivations presented in the following are especially useful for spatial encoding.

When a voltage is applied between two ends of a wire formed conductor, the Coulomb force only influences particles inside the wire. However, with two wires with electric current flow, a force between these two wires can be observed. Both wires are electrically neutral in total. Hence, this force can not be described by Coulomb's law.

To explain this phenomenon, the concept of magnetic force is used.

Like Coulomb's law is the basis for the electrostatic, Ampere's law gives the basis for the calculation of this force

$$\mathbf{F}_{12} = \frac{\mu_0 I_1 I_2}{4\pi} \oint_{C_1} \oint_{C_2} \frac{\mathbf{dr}_1 \times (\mathbf{dr}_2 \times \mathbf{r}_{12})}{r_{12}^3} \, . \tag{2.25}$$

As shown in Figure 2.3, the force emerges between two arbitrary chosen points on two conductor slopes and depends inversely on the square of the distance between them and on the current strength in each slope.

Analogue to the permittivity of vacuum ε_0 in Coulomb's law, the nature constant μ_0 is the permeability of vacuum

$$\mu_0 = 4\pi \cdot 10^{-7} \frac{\mathrm{Vs}}{\mathrm{Am}} \, . \tag{2.26}$$

This constant is connected to ε_0 by

$$\varepsilon_0 \mu_0 c^2 = 1, \tag{2.27}$$

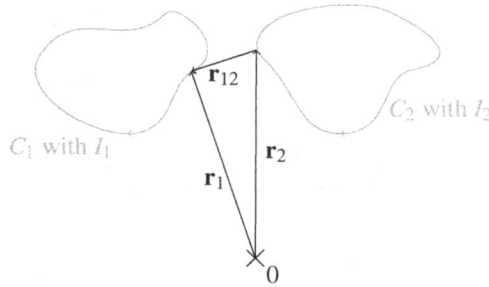

Figure 2.3: Visualization of Amperes law of magnetism. Two slopes C_1 and C_2 carry the currents I_1 and I_2. A force occurs between a point on one and a point on the other slope, being inversely proportional to the distance of the two points.

which is a result of the theory of special relativity and is not further derived here.

To be able to calculate the effect (the force) that the current in one conductor has on a second, arbitrary located conductor with an unknown current, one can split the integration of (2.25) into two steps. The first step is to calculate the inner integral

$$\mathbf{B}_2(\mathbf{r}_1) = \frac{\mu_0 I_2}{4\pi} \oint_{C_2} \frac{d\mathbf{r}_2 \times \mathbf{r}_{12}}{r_{12}^3} . \tag{2.28}$$

The resulting vector $\mathbf{B}_2(\mathbf{r}_1)$ is defined as the magnetic flux density with the unit Tesla ($T = \frac{N}{A\,m}$). The resulting force can directly be calculated by

$$\mathbf{F}_{12} = I_1 \oint_{C_1} d\mathbf{r}_1 \times \mathbf{B}_2(\mathbf{r}_1). \tag{2.29}$$

The Biot-Savart law brings the formula derived for conductor strings in a more general shape using the current density $\mathbf{j}(\mathbf{r})$

$$\mathbf{B}(\mathbf{r}_1) = \frac{\mu_0}{4\pi} \int \mathbf{j}(\mathbf{r}_2) \times \frac{\mathbf{e}_{12}}{|\mathbf{r}_1 - \mathbf{r}_2|^2} d^3 r_2 . \tag{2.30}$$

The usage of the integral in equation (2.30) implies that the summation of individual contributions to the magnetic flux densities follows the concept of superposition. This leads to the idea of magnetic fields with magnetic flux lines, which display the direction of the magnetic flux at any point in space by their tangent. Furthermore, the magnetizing field strength is characterized by the line density.

Maxwell's equations for magnetism

The spatial encoding is a crucial step for every imaging device. In MPI it is done with the help of the in the Biot-Savart law introduced superposition of magnetic fields. Therefore it is highly important, to understand the shape and density of the magnetic flux.

Starting from Biot-Savart law (2.30), the most important property of this field can be derived by reformulating the expression with the following equivalence

$$
\begin{aligned}
\mathbf{B}(\mathbf{r}_1) &= \frac{\mu_0}{4\pi} \int \mathbf{j}(\mathbf{r}_2) \times \frac{\mathbf{e}_{12}}{|\mathbf{r}_1 - \mathbf{r}_2|^2} d^3 r_2 \\
&= \frac{\mu_0}{4\pi} \int \nabla_r \times \frac{\mathbf{j}(\mathbf{r}_2)}{|\mathbf{r}_1 - \mathbf{r}_2|} d^3 r_2 \\
&= \nabla_r \times \frac{\mu_0}{4\pi} \int \frac{\mathbf{j}(\mathbf{r}_2)}{|\mathbf{r}_1 - \mathbf{r}_2|} d^3 r_2 \\
&= \mathrm{rot}\left(\frac{\mu_0}{4\pi} \int \frac{\mathbf{j}(\mathbf{r}_2)}{|\mathbf{r}_1 - \mathbf{r}_2|} d^3 r_2 \right).
\end{aligned}
\tag{2.31}
$$

This shows, that the magnetic field is a rotation field, which means that it is free of sources. A mathematical expression for this is

$$
\mathrm{div}\mathbf{B} = 0 \quad \text{or} \quad \oint_{S(V)} \mathbf{B}(\mathbf{r}) \cdot d\mathbf{f} = 0.
\tag{2.32}
$$

Both terms state, that all magnetic flux lines on the surface S of a volume V add to zero. An equivalent statement is that the start and end points of a magnetic field line are the same. This leads to the fundamental assumption, that no magnetic charges or magnetic monopoles exist, like they do for electricity. One assumes that only dipoles exist containing a north pole, where the magnetic field lines start, and a south pole, where the lines end.

With the known shape, the magnetic flux density $\mathbf{B}$ is given by

$$
\mathrm{rot}\,\mathbf{B} = \mu_0 I_{\mathrm{encl}} \quad \text{or} \quad \oint \mathbf{B}(\mathbf{r}) \cdot d\mathbf{s} = \mu_0 I_{\mathrm{encl}},
\tag{2.33}
$$

where I_{encl} is the current, enclosed by the path $\mathbf{s}$.

2.1.4 Magnetic moment and magnetization

As a preparation for the later introduced electromagnetism and for the nanoparticle model used in MPI, magnetic moments and their macroscopic representation the magnetization are introduced in the following.

A current density $\mathbf{j}(\mathbf{r})$ inside an area and the resulting magnetic induction $\mathbf{B}(\mathbf{r})$ outside this area are considered. The induction has an expression as a rotation of the vector field

$$
\mathbf{A}(\mathbf{r}) = \frac{\mu_0}{4\pi} \int \frac{\mathbf{j}(\mathbf{r}_2)}{|\mathbf{r}_1 - \mathbf{r}_2|} d^3 r_2 ,
\tag{2.34}
$$

given by equation (2.31). With a Taylor expansion of $\frac{1}{|\mathbf{r}_1-\mathbf{r}_2|}$, as the smallest not vanishing term for $\mathbf{A}(\mathbf{r})$ remains

$$\mathbf{A}(\mathbf{r}) \approx \frac{\mu_0}{4\pi} \frac{1}{2r^3} \left(\int \mathbf{r} \times \mathbf{j}(\mathbf{r}) \mathrm{d}^3 r \right) \times \mathbf{r} . \tag{2.35}$$

This motivates the introduction of the magnetic moment

$$\mathbf{m} = \frac{1}{2} \int \mathbf{r} \times \mathbf{j}(\mathbf{r}) \mathrm{d}^3 r . \tag{2.36}$$

Since the monopole contribution is zero, this moment is a dipole moment. This is an equivalent expression for the circular character of magnetic field lines.

To find a macroscopic representation of all magnetic moments inside a body with volume V at the position $\mathbf{r}$, the magnetization is introduced as

$$\mathbf{M}(\mathbf{r}) = \frac{1}{V(\mathbf{r})} \sum_{i=1}^{N(V(\mathbf{r}))} \mathbf{m}_i . \tag{2.37}$$

2.1.5   Magnetic fields in matter

With the known definition for the magnetization $\mathbf{M}$, the magnetic field

$$\mathbf{H} = \mathbf{B}\mu_0^{-1} - \mathbf{M} \tag{2.38}$$

can be introduced, featuring the same unit as the magnetization ($\frac{A}{m}$). This leads to the new expression for the magnetic flux density inside matter

$$\mathbf{B} = \mu_0(\mathbf{H} + \mathbf{M}) \tag{2.39}$$

If one now assumes a linear dependency between $\mathbf{M}$ and $\mathbf{H}$, the proportionality constant χ_m, called the magnetic suszeptibility, can be used

$$\mathbf{M} = \chi_m \mathbf{H} . \tag{2.40}$$

Its meaning becomes clear after introducing another factor, the relative permeability

$$\mu_r = 1 + \chi_m . \tag{2.41}$$

This inserted into equation (2.39) gives

$$\mathbf{B} = (1 + \chi_M)\mu_0\mathbf{H} = \mu_r\mu_0\mathbf{H}\,. \tag{2.42}$$

In terms of the density of magnetic field lines, it can be said that the density of the field lines of $\mathbf{H}$ normalized by μ_0 either increase (with $\chi_m > 0$) or decrease (with $\chi_m < 0$) in relation to the density of the field lines of $\mathbf{B}$. The easiest case is $\chi_m = 0$, which holds in vacuum

$$\mathbf{B} = \mu_0\mathbf{H}\,. \tag{2.43}$$

Chosen to be presented here are the two fundamental cases of dia- and paramagnetism and the for MPI fundamental cases of ferro- and superparamagentism. Thereby just the behavior of the magnetization and no explanation based on an atomic model is given.

Dia- and Paramagnetism

If matter is diamagnetic it is free of magnetic dipoles, if no external magnetic field is applied. Only with this application, dipoles in opposite direction to the field are induced. This means that

$$\chi_m^{\mathrm{d}} < 0 \text{ and } \chi_m^{\mathrm{d}} = \mathrm{const}\,. \tag{2.44}$$

This effect appears in any matter, but only if no other processes take place, this effect plays an important role, since the order of magnitude of χ_m is relatively small

$$\chi_m^{\mathrm{d}} \approx 10^{-5}\,.$$

The second fundamental case deals with the existence of permanent magnetic dipoles in matter. These are ordered in an external magnetic field parallel to the field lines. This order is directed against the thermodynamic tendency. These two main characteristics can be summarized by

$$\chi_m^{\mathrm{p}} > 0 \text{ and } \chi_m^{\mathrm{p}} = \chi_m^{\mathrm{p}}(T)\,, \tag{2.45}$$

where T is the absolute temperature. This behavior is called paramagnetism and dominates the effects of diamagnetism by two to three orders of magnitude

$$\chi_m^{\mathrm{d}} \approx 10^{-3...-2}\,.$$

Ferromagnetism

Ferromagnetism is a manifestation of the collective magnetism and the most important behavior of the magnetization of matter for MPI, since the signal generation relies on the non-linear magnetization curve, which is presented in the following.

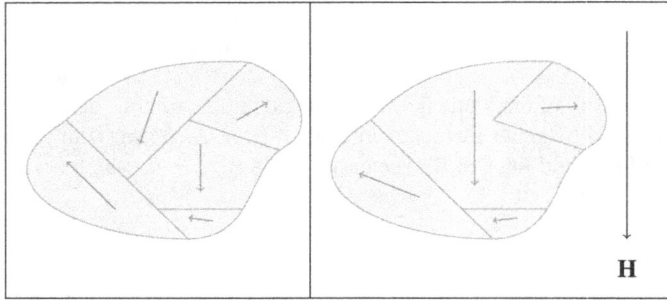

Figure 2.4: Ferromagnetic material can be modeled by elementary domains, each carrying a magnetization with an individual direction. With an applied external magnetic field **H**, the magnetization of the whole body can be changed by redirecting the magnetization of domains towards the external field and by then merging neighboring domains with a parallel direction.

This behavior is accompanied by a high magnetic susceptibility ($\chi_m^f \gg 0$). In Figure 2.4 the microscopic model for ferromagnetism is shown. It is assumed that the matter splits up into many small domains, each behaving like a volume with an own magnetization. The absolute value of the magnetic moment is thereby determined by the size of the domain. As in the case of paramagnetism, this process is dependent on the absolute temperature. The critical temperature T_C for the magnetic behavior is called the Curie-temperature. Three fundamental cases are relevant:

- $T = 0$: All magnetic moments are parallel and hence, the matter can be represented by one domain.

- $0 < T < T_C$: The for MPI important case, where the disorder increases with the temperature and a wide distribution of domain sizes and directions exist.

- $T > T_C$: The ferromagnet behaves like a paramagnet.

In the second case the magnetization of ferromagnetic matter behaves in an external magnetic field as shown in Figure 2.5. A hysteresis curve with the center at 0 T can be seen. The most important characteristic of this curve however is the non-linearity, which is different to the other presented magnetization behaviors (see Figure 2.6).

Superparamagnetism

The basis for signal generation in MPI is the non-linear magnetization curve of a tracer material. This behavior is a special case of ferromagnetism presented above. The tracer consists of a core and a shell. The core has typically a size of about $5 - 20$ nm and the shell of up to 400 nm [23]. The purpose of the shell is to prevent agglomeration of individual tracers. This is crucial for clinical applications, where blood vessels are streamed with the particles and for signal generation. Since it is obvious, how pathogen agglomeration of

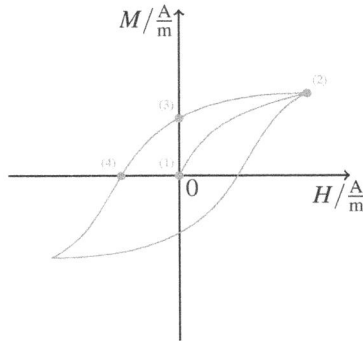

Figure 2.5: Hysteresis curve of ferromagnetic matter: At (1), the material begins to experience an external magnetic field and follows a non-linear magnetization behavior. The saturation is reached at (2). If the external field is brought to 0 again, a rest magnetization, the remanenz, stays (3), which can be compensated by a negative external field, the coerzitive force, at (4). A periodic process follows.

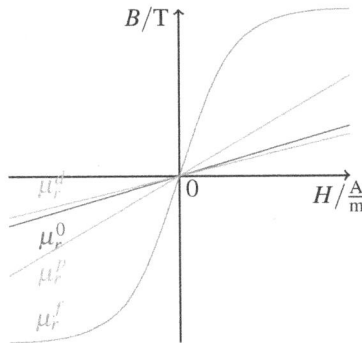

Figure 2.6: Qualitative curves, showing the factor between the magnetic field strength H and the magnetic flux density B, relative to μ_r^0. The slope μ_r^d give the relation for diamagnetic, μ_r^p for paramagnetic and the non-constant μ_r^f for ferromagnetic materials. Ferromagnetic matter with a non-linear curve are used for MPI.

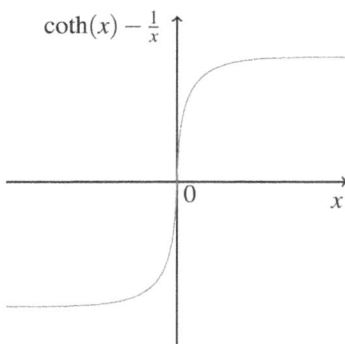

Figure 2.7: The displayed Langevin function can be used to characterize the magnetization behavior of ferromagnetic matter inside an external magnetic field.

particles inside small blood vessels are, the importance for signal generation still has to be clarified.

Multiple cores attached to each other would lose the important characteristic of these superparamagnetic iron oxides (SPIOs): Their cores consist of ferromagnetic material, such as magnetite (Fe_3O_4), but they are small enough to feature only one magnetic domain. In the above presented conception of ferromagnetism, where many magnetic domains can be controlled by external magnetic fields (Figure 2.4), the hysteresis curve emerges from single domains remaining in their size without the external magnetic field. With only one domain the hysteresis of the curve vanishes. Interaction between the cores are due to the coating not possible.

The result is a curve mathematically described by the formula

$$M(x) = M_0(x) \left[\coth(x) - \frac{1}{x} \right] . \tag{2.46}$$

The curve is displayed in Figure 2.7. A value for x and the use of this magnetization behavior is derived in the particle model in section 2.2.1.

2.1.6   Electromagnetic induction and Faraday's law

In the derivations presented so far, only static magnetic and electric fields are considered. It is already shown, that the movement of electrons induces a magnetic field. This is utilized in MPI for spatial encoding. However, the physical basics for the signal generation with magnetic particles remains to be given. The goal for imaging is to measure the change in magnetization of a distribution of SPIOs.

This is possible due to electromagnetism. When charges are accelerated inside a conductor the magnetic field, induced by the charge movement, changes over time. This leads to a change in magnetization for matter inside this magnetic field. Contrariwise, the electric field is influenced, when the magnetization changes over time. This has been found by Faraday and is called Faraday's law of induction

$$\oint_C \mathbf{E} \, d\mathbf{r} = -\frac{d}{dt} \int_{F_C} \mathbf{B} \, df, \qquad (2.47)$$

where C is a closed circuit slope.

The differential equivalent for formula (2.47) can be derived into

$$\mathrm{rot}\, \mathbf{E} = -\dot{\mathbf{B}} \qquad (2.48)$$

by using Stokes' theorem. Hence, the change in the magnetic field causes a rotation in the electric field.

With this, the fundamentals in electrostatic, magnetostatic and electrodynamics are given. In the following section one can directly see the central role of the given laws in MPI. Thereby, the Biot-Savart law and the concept of magnetic fields are used to understand the coil setup and the effect on the nanoparticles. The nanoparticles themselves are explained by the magnetic moment and the laws of superparamagnetism, which in turn relies on the theory of ferromagnetism. The theory for the receiving of signals uses Faraday's law of induction. The section about electric networks makes especially use of the firstly presented theory of electrostatic.

2.2 Magnetic particle imaging

An imaging system needs two crucial qualities: It has to be able to detect a signal from an area of interest and it has to provide a spatial encoding inside this area. In MPI the signal generating entities are magnetic nanoparticles, the superparamagnetic iron oxides (SPIOs) [1]. Their particle response to an external magnetic field can be recorded. With the simple superposition of a second magnetic field, the selection field (SF), a signal generation only in a specified area is possible. Therefore a combination of the signal generating drive field and the SF offers spatial encoding.

2.2.1 The particle model

It is helpful to first understand the behavior of the nanoparticles without any spatial encoding. Their nonlinear magnetization characteristic in an external magnetic field can be utilized to create a signal, which is proportional to the particle concentration.

The magnetization of the particles is described by the Langevin function, which has already been presented in equation (2.46). The resulting behavior of the magnetziation over time is

$$\mathbf{M}(\mathbf{r},t) = M_0(\mathbf{r}) \left[\coth\left(\frac{m|\mathbf{B}(\mathbf{r},t)|}{k_B T} \right) - \frac{k_B T}{m|\mathbf{B}(\mathbf{r},t)|} \right] \mathbf{e_B}(\mathbf{r},t) . \tag{2.49}$$

Here $\mathbf{B}(\mathbf{r},t)$ is the external magnetic field, whose direction is given by $\mathbf{e_B}(\mathbf{r},t)$, k_B is the Boltzmann-constant, T is the absolute temperature and m is the magnetic moment of a single particle (see section 2.1.4). Furthermore, $M_0(\mathbf{r})$ is the saturation magnetization of a voxel with size ΔV and can be expressed as

$$M_0(\mathbf{r}) = \frac{N(\mathbf{r})\overline{m}}{\Delta V} , \tag{2.50}$$

where $N(\mathbf{r})$ is the number of particles inside the voxel [24]. It holds that the mean magnetic moment of each particle $\overline{m} = \frac{1}{6}\pi d^3 M_s$ with a saturation magnetization of magnetite of $M_s = 0.6\ \mathrm{T}\mu_0^{-1}$ [25].

2.2.2   Signal generation

With the described particle model, the signal generation can easily be achieved by alternating the external magnetic field $H(t)$ sinussoidally around 0 Tesla

$$H(t) = H_0 \sin(t f_0) . \tag{2.51}$$

If H_0 is chosen high enough to achieve a saturation of the particle magnetization, a signal proportional to the particle concentration can be achieved. The proportionality to the concentration has already been expressed in the particle model (see equation (2.50)), where the saturation magnetization M_0 scales with $N(\mathbf{r})$, the number of particles at a certain position $\mathbf{r}$.

The signal is generated by the change of magnetization of the particles, which leads to an induction of voltage proportional to this change in a receiving coil, placed in proximity to the particles. A spectrum of the resulting signal, induced by the change in magnetization over time consists not only of the fundamental frequency f_0, but also of particle specific frequencies, as derived below.

A mathematical derivation can be achieved by inserting the sinusoidal external magnetic field into the particle model, which is done here in one dimension

$$M(x,t) = M_0(x) \left[\coth(\gamma \sin(t f_0)) - (\gamma \sin(t f_0))^{-1} \right] . \tag{2.52}$$

For the sake of clarity $\gamma = \frac{m\mu_0 H_0}{k_B T}$ is used. The result is an alternating and odd function. The measurement only allows to detect the change of the magnetization, following the law of induction. Mathematically this means that the time derivative of the function 2.52 is detected

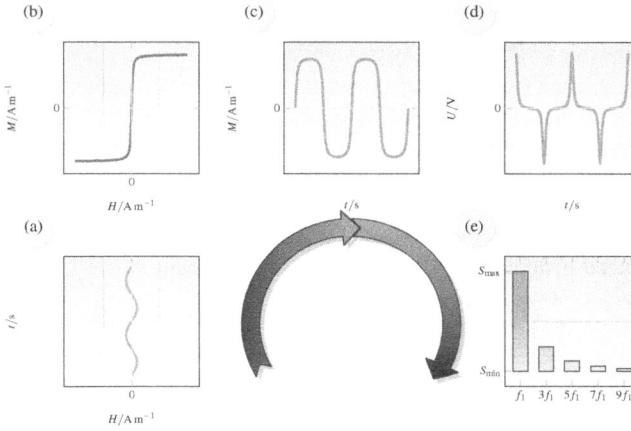

Figure 2.8: The principle of signal generation in MPI is displayed. In (a) the sinussoidal external magnetic field is given. (b) shows the particles magneti- zation curve, following the Langevin function. When applying the external field to this function, the particle magnetization over time is no pure sine, but a smoothed rectangular function (c). Measurable is only the change in magnetiza- tion (d). The Fourier spectrum of the measured signal (e) contains beside the fundamental frequency also higher odd harmonics of it. Graphic adapted from [21].

$$\frac{\partial}{\partial t}M(x,t) = M_0 \left[\gamma^{-1}\cot(t)\csc(t) - \gamma\cos(t)\sinh^{-2}(\gamma\sin(t)) \right] . \tag{2.53}$$

This is the general shape of the signal, detected in the receiving coils and saved on the data carrier. Here, the coil sensitivity of the receiving coils still influences this signal as well as the total volume, the particle concentration is in. This is further analyzed in section 2.2.5 and has been omitted here.

The function is even, but again alternating. Through its alternating nature, the even Fourier coefficients of this function vanish. Therefore the signal spectrum consists not only of the fundamental frequency f_0, but also of odd higher harmonics of this frequency: $3f_0 = f_3, 5f_0 = f_5, 7f_0 = f_7$, etc.. The amplitude of these coefficients is proportional to the particle concentration and through reference value, the particle concentration can be measured.

It is also notable that a signal encoding without using the frequency spectrum is possible. This approach is called x-space imaging and can be found in [26, 27].

2.2.3   Spatial encoding

In section 2.2.1 it is described that it is possible to determine, whether or not particles are in a magnetic field. If a heterogeneous particle distribution is in the field of interest and one wants to find, where the particles are and how the distribution looks like, spatial encoding is needed.

In MPI this is achieved by applying an additional external magnetic field, the SF $\mathbf{H}_{SF}$. The purpose of this field is to generate a signal only in a specified area FOV. Again, the magnetization curve of the SPIOs is important. As shown in Figure 2.9, a magnetic field high enough to bring the magnetization of the nanoparticles into saturation can be superimposed. This means, that an oscillating external magnetic field would have only a minor effect on the change of magnetization. A minor change of magnetization over time means, that almost no signal is induced inside the receiving coils. However, where $\mathbf{H}_{SF}$ equals zero, a typical signal generation as shown in section 2.2.2 happens.

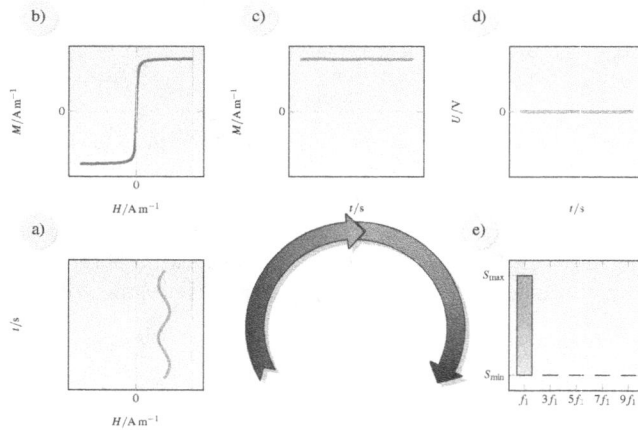

Figure 2.9: The spatial encoding mechanism is visualized. The encoding is achieved by applying an offset field to the oscillating DF (a). This brings the particle magnetization into saturation (b). This way, the sinussoidal field changes the particle magnetization only marginally (c) and the time derivative of this magnetization curve stays close to zero (d). Hence, no signal is detected with the exception of the fundamental frequency that comes directly from the oscillating external magnetic field (e). Graphic adapted from [21].

One approach to form the SF is to generate a field that applies the saturation case to all particles in the FOV except for one point. This so called field free point (FFP) is then the spot for signal generation [28]. Furthermore, it is possible to steer this point over the full ROI by a drive field. If now the particles inside a voxel of the FOV are in saturation due to the SF and the FFP passes by, their magnetization is changed with the frequency of the drive field. This exactly fulfills the conditions for signal generation as presented in section 2.2.2.

Recently, another shape for the SF has been presented [14]. With a field free line (FFL) in a two dimensional FOV, a spatial encoding scheme similar to that of computed tomography can be used [29]. The SF with an FFL ideally generates no magnetic field directly on a straight line and features a homogeneous gradient perpendicular to the center line. Here, more particles experience an external magnetic field of zero Tesla and hence generate a signal at the same time (all particles on the line). The signal is strengthened and consequently

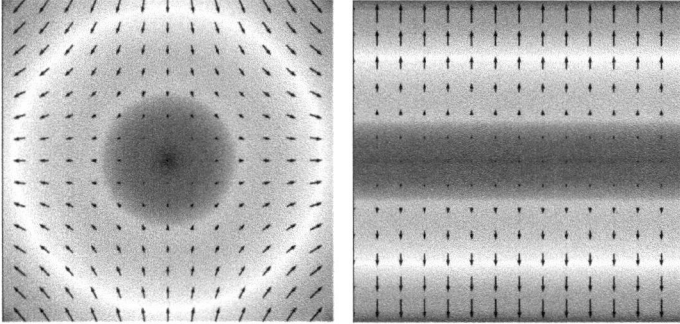

Figure 2.10: FFP and FFL pic mit spulen drum rum.

the SNR is increased. Furthermore, a faster scan due to a less dense trajectory is possible (see 2.2.4).

This shape of $\mathbf{H}_{SF}$ can be achieved by placing three or more coil pairs in Maxwell configuration on a circle around the FOV with the main axis of each coil pointing towards the center of the FOV [15]. As suggested in [15], four coil pairs are chosen for an optimal field in the xy-plane, where the perpendicular z-axis is the main axis of the circle, the coils are positioned on.

2.2.4   Reconstruction in MPI

With the spatial encoding scheme using an FFL and the signal generation at the desired spots, an image has to be reconstructed from the raw data. To understand this, the generated fields have to be described mathematically.

A magnetic field with an FFL along the by γ relatively to the laboratory coordinate system rotated direction $\mathbf{d}_{\mathrm{FFL}}^{\gamma} = (\cos\gamma, \sin\gamma, 0)^T$ with an ideal coil setup is given by

$$\mathbf{H}_S^{\gamma}(x,y) = (Gx\sin\gamma - Gy\cos\gamma) \cdot \begin{pmatrix} -\sin\gamma \\ \cos\gamma \\ 0 \end{pmatrix} , \qquad (2.54)$$

with G being the gradient strength of the field [15]. Here, the z-direction is neglected, and only the xy-plane is considered. It can easily be seen that the direction of this field is perpendicular to the above introduced line $\mathbf{d}_{\mathrm{FFL}}^{\gamma}$ and constant parallel to it. Hence the SF is perfectly described. For the calculations round coils with an infinite circle diameter are assumed.

To acquire enough information for reconstruction, the field is moved over the xy-plane to full extend with the drive field

$$\mathbf{H}_D^\gamma(t) = A\Lambda(t) \begin{pmatrix} -\sin\gamma \\ \cos\gamma \\ 0 \end{pmatrix}.$$
(2.55)

Thereby $\Lambda(t)$ is the sinussoidal excitation function $\cos(2\pi f_0 t)$ and A is the field amplitude. The total magnetic field adds to

$$
\begin{aligned}
\mathbf{H}^\gamma(x,y,t) &= \mathbf{H}_S^\gamma(x,y) + \mathbf{H}_D^\gamma(t) \\
&= (A\Lambda(t) + Gx\sin\gamma - Gy\cos\gamma) \cdot \begin{pmatrix} -\sin\gamma \\ \cos\gamma \\ 0 \end{pmatrix}.
\end{aligned}
$$
(2.56)

This field translates and rotates the FFL through the FOV [17]. Thereby the displacement of the FFL caused by the drive field is given by

$$\xi_{\text{FFL}} = \frac{A}{G}\Lambda(t).$$
(2.57)

2.2.5   Raw data

As shown in the particle model in section 2.2.1, the particles change their magnetization with the external field. The magnetization **M** from this section can also be expressed with the desired distribution of nanoparticle concentration $c(x,y)$:

$$\mathbf{M}(x,y,t) = \frac{N}{\Delta V}\overline{\mathbf{m}}(x,y,t), = c(x,y)\overline{\mathbf{m}}(x,y,t)$$
(2.58)

Here $\overline{\mathbf{m}}$ is the mean magnetic moment that can be calculated by the Langevin function 2.49.
As shown in [17], the rotational symmetry of $\overline{\mathbf{m}}$ yields

$$\overline{\mathbf{m}}(x,y,t) = \overline{m}(A\Lambda(t) + Gx\sin\gamma - Gy\cos\gamma) \cdot \begin{pmatrix} -\sin\gamma \\ \cos\gamma \\ 0 \end{pmatrix}.$$
(2.59)

The raw data is the induced signal in the receiving coils (with sensitivity $\mathbf{p}(x,y)$), following the law of induction (see 2.47) and is given by

$$
\begin{aligned}
u^\gamma(t) &= -\mu_0 \int_{\mathbb{R}^2} \frac{\partial}{\partial t}\mathbf{M}(x,y,t) \cdot \mathbf{p}(x,y)\ \mathrm{d}x\mathrm{d}y \\
&= -\mu_0 \int_{\mathbb{R}^2} c(x,y)\frac{\partial}{\partial t}\overline{\mathbf{m}}(x,y,t) \cdot \mathbf{p}(x,y)\ \mathrm{d}x\mathrm{d}y
\end{aligned}
$$
(2.60)

2.2.6 Reconstruction step

Knowing the raw data, the reconstruction step can be done by using the theory of radon space [30]. In [17], it is shown that the signal u^γ in (2.60) is a convoluted radon transform $\mathcal{R}(c)$ of the desired image $c(x,y)$

$$u^\gamma(t) = A\Lambda'(t)(\tilde{m} * \mathcal{R}(c)(\gamma, \cdot)) \left(\frac{A}{G}\Lambda(t) \right) . \tag{2.61}$$

The Radon transformation is defined as

$$\mathcal{R}(c)(\gamma, \xi) = \int_{\mathbb{R}} c(v\cos\gamma - \xi\sin\gamma, v\sin\gamma + \xi\cos\gamma)dv . \tag{2.62}$$

The convolution kernel results from the derivative of the particles' mean magnetic moment

$$\tilde{m}(x) = -\mu_0\overline{m}(Gx) . \tag{2.63}$$

To be able to perform an inverse Radon transformation to acquire the desired particle distribution, the Radon transformed of the distribution first has to be isolated.

For the given considerations the receive channel normalization, which can be found in [17], is left out. However, it is notable that the normalization yields that two, non-parallel receive coils are needed for a full reconstruction.

Since the FFL varies its speed when changing the direction, a speed normalization is required:

$$s^\gamma(t) = \frac{\tilde{u}^\gamma}{A\Lambda'(t)} = (\tilde{m} * \mathcal{R}(c)(\gamma, \cdot)) \left(\frac{A}{G}\Lambda(t) \right) . \tag{2.64}$$

Now the time dependent signal has to be made spatial dependent by the transformation, prepared in equation (2.57). The interval for the displacement of the FFL to cover the full area of interest is $[-\frac{A}{G}, \frac{A}{G}]$. The transformation is given by

$$\tilde{s}^\gamma(\xi_{\text{FFL}}) = \tilde{s}^\gamma \left(\Lambda^{-1} \left(\frac{G}{A}\xi_{\text{FFL}} \right) \right) . \tag{2.65}$$

Now the radon data is isolated:

$$\mathcal{R}(c)(\gamma, \xi) = \mathscr{F}^{-1} \left(\frac{\hat{s}^\gamma(v)}{\hat{m}(v)} \right) . \tag{2.66}$$

Here, $\hat{s}^\gamma(v)$ and $\hat{m}(v)$ are the Fourier transform of $\tilde{s}^\gamma(\xi)$ and $\tilde{m}(\xi)$.

To conclude, the reconstruction for MPI using an FFL can be given by

$$c(x,y) = \mathscr{R}^{-1} \left(\mathscr{F}^{-1} \left(\frac{\hat{s}^\gamma(v)}{\hat{m}(v)} \right) \right). \tag{2.67}$$

The actual inverse radon transformation can be achieved by known reconstruction techniques from computed tomography, like the filtered backprojection, ARS or filtered layergram [29] and could already be performed in practice [31].

Notable is also the reconstruction based on a system matrix [32–39]. The presented derivations show that reconstruction using an FFL is possible. A reconstruction approach with a system matrix will be used for first results, but a comparison with a Radon based reconstruction is prospective.

A system matrix based reconstruction premises a recording of the scanners system matrix. A generation of a system matrix can be found in [32]. Then a system of equations has to be solved to reconstruct the particle distribution

$$\mathbf{S} \cdot \mathbf{c} \approx \mathbf{u}. \tag{2.68}$$

Here, $\mathbf{S}$ is the recorded system matrix, $\mathbf{u}$ is the recorded signal and $\mathbf{c}$ is the particle concentration distribution, which is calculated by solving the system. Since the signal is noisy, the solution of the system is only an approximation to a perfect solution of the reconstruction problem.

2.3   Electric Networks

The central part of this work is the signal chain, that sends the computer generated signal to the transmission coils in the MPI scanner and the signal chain, that carries the recorded raw data from the receiving coils back to the computer. In this section, the components used in these networks are introduced and their interaction is examined. The main focus is set on hardware filtering.

2.3.1   Fundamental dipoles

The three fundamental components for an electric network are the resistance R, the inductivity L and the capacity C.

Resistance

The simplest element is the resistance. Starting with the assumption of linearity given by equation (2.6), one can see that

$$U = \int_{(R)} \frac{\mathbf{F}}{q} \, d\mathbf{r} = \sigma \int_{(R)} \mathbf{j} \, d\mathbf{r} = \sigma \frac{l}{F} I, \tag{2.69}$$

where σ is the proportionality constant and l and F are the length and cross section of the resistor. The material constant σ is called specific resistance and has the unit Ωm. Hence for the resistor it holds that

$$U_R = IR. \tag{2.70}$$

Capacitor

When two parallel conductive plates with areas F are brought into close proximity d, the system and both plates feature an equal charge Q, the system is called a capacitor. The resulting electric field features a non zero value in between the plates of

$$\mathbf{E}(\mathbf{r}) = \frac{Q}{F\varepsilon_0}\mathbf{e}_n, \tag{2.71}$$

where $\mathbf{e}_n$ is the normal vector on the negative charged plate. The corresponding electrostatic potential is

$$\varphi(\mathbf{r}) = \frac{-Q}{F\varepsilon}z. \tag{2.72}$$

Here, z is a distance from one plate to a point in between the two plates ($0 < z < d$). The applied voltage is

$$U = \varphi(z=0) - \varphi(z=d) = \frac{Q}{F\varepsilon_0}d. \tag{2.73}$$

Hence, we obtain a linear proportionality between the total charge of the capacity and the voltage

$$U_C = \frac{Q}{C}, \tag{2.74}$$

with $C = \varepsilon_0^{-1}F^{-1}$ being the capacitance of the system with the unit Farad ($[C] = F = \frac{As}{V}$).

Inductor

The third fundamental component is the inductor. If one assumes a constant permeability around a conductor loop, Maxwell's equations for magnetism state that the magnetic flux density $\mathbf{B}$ is proportional to the flowing current in the conductor (see equation 2.32). Hence the magnetic flux

$$\phi = \mathbf{B} \cdot \mathbf{A} \tag{2.75}$$

in the area **A** has a linear dependency on the current

$$\phi = LI.$$ (2.76)

The proportionality constant L is the inductance of the current loop and has the unit Henry ($[L] = H = \frac{Vs}{A}$). For a coil with N windings it holds

$$\phi = \frac{LI}{N}.$$ (2.77)

To find the dependency of the voltage applied to such a coil , the current and the inductance, Faraday's law of induction (2.47) can be used

$$
\begin{aligned}
U &= \oint \mathbf{E}_i \mathrm{d}\mathbf{s} \\
&= -N - \frac{\mathrm{d}\phi}{\mathrm{d}t} \\
&= -\frac{\mathrm{d}LI}{\mathrm{d}t} \\
&= L\frac{\mathrm{d}I}{\mathrm{d}t} + I\frac{\mathrm{d}L}{\mathrm{d}t} \\
&= L\frac{\mathrm{d}I}{\mathrm{d}t}.
\end{aligned}
$$ (2.78)

The last step acts on the assumption that the inductance L does not change over time, which is true for the presented cases.

In summary for a coil it holds

$$U_L = -L\dot{I}.$$ (2.79)

2.3.2   Kirchhoff's laws

When conductor slopes are connected and a current flows, two fundamental laws can be derived, which establish the basis for electric network analysis. For the definition of node and loops see Figure 2.11.

Kirchhoff's first law states about the sum of all currents in one node

$$\sum_{j=1}^{n} I_j = 0.$$ (2.80)

This is a direct application of the continuity equation (2.8) to electric conductors.

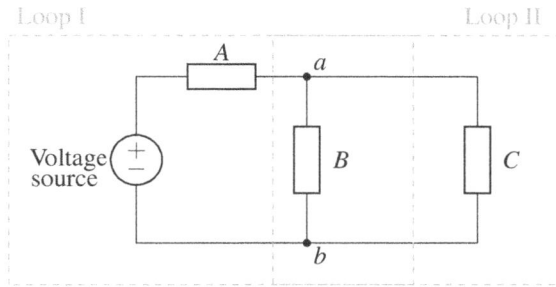

Figure 2.11: An electric circuit with dipoles A and B positioned in Loop I and B and C in Loop II. Connections between loops are called nodes and are marked by a and b.

In Kirchhoff's second law, it is stated, that the sum of all voltages in one loop equals

$$\sum_{loop} U_j = 0 \,. \tag{2.81}$$

The second rule emerges from the fact that electric potentials are conservative, which in turn is a result of equation (2.20).

Note that for the correct use of these laws, the algebraic signs of voltages at sources and loads have to be contrariwise.

2.3.3 Interaction of network components

With the known behavior of single dipoles with flowing alternating current and the knowledge about Kirchhoff's laws, a network, which contains all components, called oscillating circuit as shown in Figure 2.12 can be analyzed.

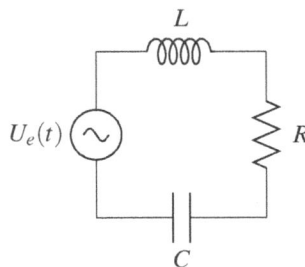

Figure 2.12: An oscillation circuit with a voltage source $U_e(t)$, an inductance L, a resistance R and a capacity C.

Combining equations (2.70), (2.74) and (2.79) results in

$$U_e = U_R + U_C + U_L = IR + \frac{Q}{C} + \dot{I}L.$$ (2.82)

With equation (2.5) and by taking the time derivative, the differential equation

$$\dot{U}_e = \frac{I}{C} + \dot{I}R + \ddot{I}L$$ (2.83)

can be obtained.

With a periodic voltage of the shape

$$U_e = U_0 \cos \omega t$$ (2.84)

the differential equation

$$-U_0 \omega \sin \omega t = \frac{I}{C} + \dot{I}R + \ddot{I}L$$ (2.85)

has to be solved. The calculation becomes easier, when the sine is replaced with an exponential function

$$U_e = U_0 e^{i\omega t}.$$ (2.86)

This leads to the solution of (2.83)

$$I(t) = I_0 e^{i(\omega t - \varphi)}.$$ (2.87)

The introduction of the exponential function yields a complex contribution to current and voltage. Physically measured is only the real part of these quantities

$$\begin{aligned} \mathrm{Re}\, I(t) &= I_0 \cos(\omega t - \varphi) \text{ and} \\ \mathrm{Re}\, U_e(t) &= U_0 \cos(\omega t). \end{aligned}$$

With the introduction of a complex current and voltage, the resistance can be redefined to

$$Z = \frac{U}{I} = \frac{U_0}{I_0} e^{i\varphi} = |Z| e^{i\varphi}$$ (2.88)

and is called complex resistance. The single parts of (2.88) are called

- Impedance $|Z| = \frac{U_0}{I_0} = \sqrt{(\mathrm{Re}\, Z)^2 + (\mathrm{Im}\, Z)^2}$,

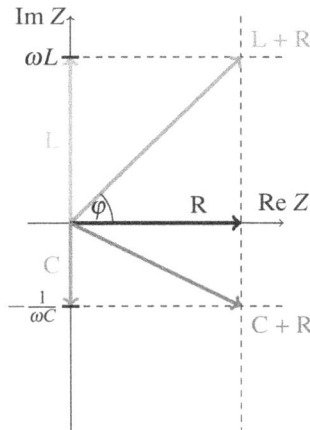

Figure 2.13: The complex resistance for different cases are displayed with arrows, where the arrow length represents impedance. The cases are from top to bottom: A pure inductance, a combination of inductance and resistance, a pure resistance, a combination of capacity and resistance and a pure capacity.

- Active resistance $\operatorname{Re} Z$,

- Reactance $\operatorname{Im} Z$ and

- Phase shift $\varphi = \arctan\left(\frac{\operatorname{Im} Z}{\operatorname{Re} Z}\right)$.

To complete the concept formation, the admittance is introduced as

$$Y = Z^{-1}. \tag{2.89}$$

2.3.4 Hardware filtering

The aim of this work is to present the transmission and receiving chain of an MPI scanner. A central challenge is thereby to build electronic filters that either damp disturb signals on the transmission side or damp the base frequency on the receiving side. Of the values of an oscillating circuit the two important parameters for this task are the phase shift and the impedance. These are the parameters that are further analyzed. Further, all implemented filters consist of individual levels made of parallel or serial circuits of a capacity and an inductance. Hence, these two cases are analyzed with impedance and phase shift.

With the complex values introduced in equation (2.86) one can write for the serial network shown in Figure 2.14 that consists of a capacity C and a inductance L (LC-network)

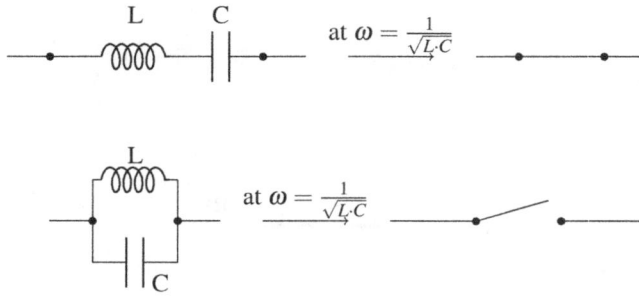

Figure 2.14: A serial resonant circuit acts like a perfect closed circuit at resonance, whereas a parallel resonance circuit acts like a perfect open circuit.

$$Z_s = i\omega L + \frac{1}{i\omega C}$$

$$= i\left(\frac{L}{\omega}\right)\left(\omega^2 - \frac{1}{LC}\right). \tag{2.90}$$

The impedance is zero when the angular frequency is $\frac{1}{\sqrt{LC}}$ and the setup can be considered as a perfect short circuit [40]. Analogue to the serial case, the ideal parallel case has a resonant frequency of $\frac{1}{\sqrt{LC}}$, since

$$Y_p = i\omega C + \frac{1}{i\omega L}$$

$$= i\left(\frac{L}{\omega}\right)\left(\omega^2 - \frac{1}{LC}\right). \tag{2.91}$$

The important difference is that at the resonant angular frequency a perfect open circuit results.

When the exact maximum or minimum value of the impedance is unknown, a reliable quantity to build a circuit with a desired resonant frequency f_0 is the phase shift. The complex representation of the resistance in Figure 2.13 shows that at the resonant case in a parallel and serial LC-network

$$\omega L = \frac{1}{\omega C} \tag{2.92}$$

the reactance $\operatorname{Im} Z$ becomes zero. This means for the phase shift

$$\varphi = \arctan \frac{\operatorname{Im} Z}{\operatorname{Re} Z} = \arctan 0 = 0. \tag{2.93}$$

This value is used for this work to find the resonant cases for all filters. In addition to this it is of interest, what happens with currents with other frequencies than f_0.

Non-resonant frequencies

At the example of a serial network (see Figure 2.14), impedance behavior at different frequencies can be analyzed. The frequency curve of the impedance can be calculated by equation (2.90) and is shown in Figure 2.15. Here, one can see that signals with other frequencies than f_0 experience a non-zero impedance with a quadratic trend. Following Ohm's law given in equation (2.21), the amplitude of the currents with other frequencies are damped linearly to this impedance. This effect can be used to damp certain frequencies, which is called filtering in electronic networks.

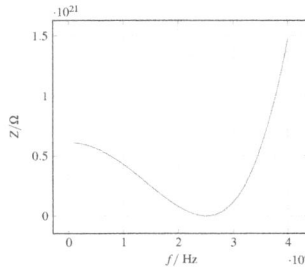

Figure 2.15: Impedance curve for a serial oscillating circuit with resonance at a frequency of 25kHz. The impedance increases quadratically with distance to the resonance case.

In the terminology of filters one can call a serial resonant circuit a bandpass filter, since it provides no impedance at resonance. A parallel resonant circuit can be called a bandstop filter, since it provides maximum impedance at resonance. Further the resonant frequency can be called the center frequency.

Since it is possible to design filters, that feature a bandpass or bandstop spectrum with more than one frequency, the definition of bandwidth of a filter is also appropriate. This is the spectral width around the center frequency, where a damping of under 3 dB occurs.

To achieve a desired damping at chosen frequencies and an overall desired behavior of the filter, it is possible to combine several levels of the presented LC-networks.

2.4 Shielding

In MPI electromagnetic induction inside the receiving coils caused by the magnetic nanoparticles is the desired signal. This means that all other sources of electromagnetic waves, that could reach the receiving coil have to be damped since they would induce false signals. The best way of doing this is shielding. The physical basis for shielding is the skin effect, which can be phenomenological be described with the basics in electromagnetism introduced in section 2.1.6.

2.4.1   Skin effect

If considering a section of a conductor wire with applied alternating current as introduced in equation (2.84), the skin effect explains the non homogeneous current distribution inside the section. The starting point is the change of the current direction in a sinussoidal manner.

This change induces a magnetic field around and also inside the conductor as shown in Figure 2.16, following Ampere's law of induction (2.25). With the change of the current direction, also the direction of the induced magnetic field changes and hence induces eddy currents around the magnetic field lines, following Faraday's circuit law (2.47). These eddy currents are directed antiparallel to the current direction inside the magnetic field lines and hence closer to the conductor center. The resulting effect is, that a countervoltage is induced, that increases in strength with the proximity to the conductor center. Consequently, the current density close to the center is reduced.

This exponential progress follows the law

$$J(d) = J_s e^{-\frac{d}{\delta}}, \tag{2.94}$$

where $J(d)$ is the current density, d is the penetration depth, J_s is the surface current density, and

$$\delta = \sqrt{\frac{2\rho}{\omega \mu_0 \mu_r}}. \tag{2.95}$$

Here ω is the angular frequency of the change in voltage, μ_0 and μ_r are the permeability of vacuum and the conductor, and ρ is the specific resistance (see (2.69)).

The same effect occurs, when electromagnetic waves enter a conducting material. The alternating electric field of the wave has an analog effect to the alternating electric voltage used for equation 2.94. Hence, the depth of penetration until e^{-1} of the wave amplitude is reached equals δ. This is used to shield individual components in 3.6.1.

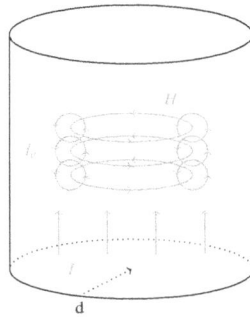

Figure 2.16: Visualization of the skin effect in a cylindrical conductor: An alternating current I induces a circular magnetic field H inside the conductor. H again induces an electric field around its field lines and since the electric field is in a conductor, it causes a current, called eddy current I_e. This is antiparallel to the applied current I in the center and parallel at the outside. Since H gets denser in the middle, the counter currents in the center of the conductor are stronger and the current density becomes smaller close to the center and higher on the surface.

Chapter 3

Materials and Methods

To understand the profitableness of the effort that is invested in a possibly clean transmission signal and in shielding of individual parts, first the scanner buildup is described in detail. After this, the signal chain on the transmission side is explained. Here important aspects are the prevention and the inhibition of harmonics of the transmission frequency f_0. The particle response will be recorded in the receiving coils, which are followed by the receiving signal chain. The purpose of this signal chain is to carry the particle response to the data medium. Here the transmission frequency f_0 has to be disabled, while influencing the particle response to a minimum. In the end, it is shown, how the induction of alternating current in the selection field coils with direct current is prevented.

3.1   Coil and magnet assembly

The field generating part of the scanner consists of three basic components

- the drive field coil assembly

- the selection field coil assembly

- two permanent magnets.

A gateway in the middle of the scanner with a bore diameter of 38 mm allows the insertion of probes. The symmetry axis of this tube is considered as the main axis of the scanner and the z-axis in the cartesian coordinate system. Examples for possible probes are a SPIO delta probe, a SPIO phantom and living mice with injected SPIOs.

3.1.1   Generation of a rotatable field free line

The principle producibility of a rotatable FFL with coil pairs in Maxwell configuration [41] has been shown in [14]. Here 16 coil pairs have been used. After this it could mathematically be proven that the creation of a rotatable FFL with three or more coil pairs is feasible [15]. For the scanner presented here, a setup with four coil pairs has been favored over a configuration with three or more than four Maxwell coil pairs, since the field quality in the xy-plane is most important for the planned imaging and this is reached with four coil pairs. These two setups and the next generation are shown in Figure 3.1.

The efficiency of this setup can be further increased [42] as presented in Figure 3.1. The two coil pairs can be arranged with different distances to the main axis. Thereby the diameter of the individual coils can be increased, which leads to an improved field quality at the same power loss. A dedicated Maxwell coil pair orientated in z-direction can be used to generate the static part of the FFL [16]. Since these coils are supplied with direct current and hence generate a static field, they can be replaced by permanent magnets, arranged in opposite directions.

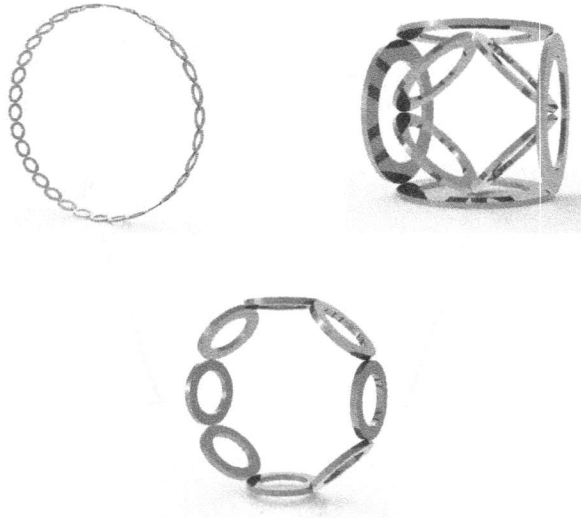

Figure 3.1: Coil setups, feasible of generating an FFL. Following the arrows, the first generation proposed 16 Maxwell coil pairs, the second 4 coil pairs and the third generation changes the coil size and topology again to achieve the lowest power loss of the three presented setups. The last topology is used in this work.

After the topology of the coils was set, the coil form could be optimized by using curved rectangular coils [43–45] instead of plain circular coils (see Figure 3.2). With the cylindrical character of the scanner, the new coils provide a better distribution of the current density. Consequently a lower power loss and an improved field quality can be achieved again. Hereby a low field strength (below 100μ T) along and a homogeneous gradient perpendicular to the FFL is considered as high field quality.

The above presented scanner topology has been used in the scanner presented in this thesis and is presented in [46]. To find the current density for an acceptable field quality, an optimization progress with a regularized least squares method was used. Thereby the distance

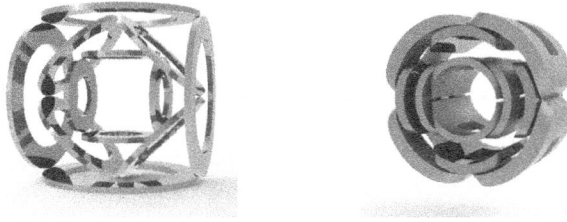

Figure 3.2: Both coil setups are feasible to generate and rotate an FFL (outer two coil layers) and to steer the FFL over the FOV (inner coils). Changing the morphology from circular coils to curved rectangular coils provides a better distribution of the current density and hence a reduced power loss at constant field quality.

$\mathcal{M}$ between an optimal target field and a simulated field of the scanner was minimized, while a penalty term $\mathcal{P}$, which represents the power loss, prevents a too high current density in the inner and outer SF coils

$$\mathcal{M}(I_i, I_o) + \mathcal{P}_{\lambda,\tau}(I_i, I_o) \overset{!}{=} min. \tag{3.1}$$

While the currents I_i in the inner coils and I_o in the outer coils are to be found, the parameters λ and τ are adjustable to find a practical solution while providing a field quality within the set constraints.

The complete term shows, how the power loss is calculated and weighted and how the difference of the two fields is computed

$$||\mathbf{H}_t - (\mathbf{h}_{s,inner} \cdot \mathbf{I}_i + \mathbf{h}_{s,outer} \cdot \mathbf{I}_o)||_2^2 + \lambda^2 \cdot \mathbf{I}_i^2 \cdot \mathbf{R}_i + \tau^2 \cdot \mathbf{I}_o^2 \cdot \mathbf{R}_o \overset{!}{=} min. \tag{3.2}$$

Here $\mathbf{H}_t$ is the desired target magnetic field in the FOV, $\mathbf{h}_s$ is the field profile of the scanner in the FOV and can be scaled to a magnetic field by multiplying the field of a single SF generating coil with the applied current. Furthermore $\mathbf{R}_i$ and $\mathbf{R}_o$ are the resistances of the inner and outer selection field coils.

To provide the desired field strength with a good cooling access, the coils were split into stacks with a 1 mm distance between each component. The final coil assembly consists of the gateway, 4 DF coils, 8 inner SF coils and 12 outer SF coils, ash shown in Figure 3.3.

To provide the static part of the FFL field, two permanent magnets with opposite direction have been chosen. This lowers the power loss of the scanner and provides a good field quality. Their installation can be found in [21].

3.1.2 Translation of the FFL and signal generation

The purpose of the DF is to translate the FFL over the region of interest in a direction perpendicular to the line. A great advantage of MPI is, that this movement allows the spatial

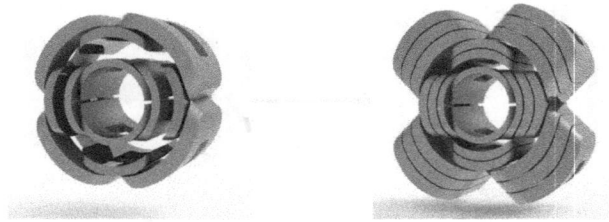

Figure 3.3: To achieve the desired field strength and field quality of the FFL, coils were stacked. This also provides a good accessibility for cooling.

encoding and the signal generation at the same time, as presented in section 2.2.

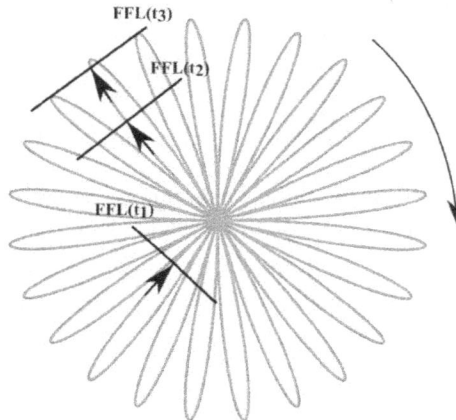

Figure 3.4: Circular trajectory of the FFLs center. The outer arrow represents the FFL rotation. The FFL is steered perpendicular to the line over the ROI.

This signal generation is executed at a base frequency of $f_0 = 25$ kHz. The FFL is moved across the FOV at this high frequency and rotated by the SF coils at a low rotation frequency (f_{rot}). The resulting shape of the trajectory is given in Figure 3.4. The ratio

$$P = \frac{f_0}{f_{rot}} \tag{3.3}$$

determines the density of the trajectory and consequently the resolution. In this scanner a rotation frequency of $f_{rot} = 100$ Hz is desired.

The strong oscillations of the DF might cause eddy currents inside the SF coils and the power consumption of the DF coils, when placed outside the SF coils, can not be cooled by air. Therefore the DF-coils were placed directly on the gateway and are then followed by the SF assembly.

An opening angle for the DF coils of 120 deg, as shown in figure 3.5, provides a reliable field translation without a too high power loss [47].

The placement of the SF coils outside the DF-coils means a higher current density in the FFL generating part. That problem could be solved by an optimization algorithm between field quality and current density that is presented in 3.1.1.

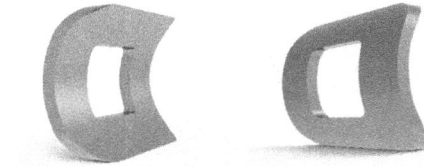

Figure 3.5: Design of the DF coils with an opening angle of 120°.

3.1.3   Installation and receiving coils

The installation of the coil assembly has been performed in the master thesis [21]. Here the principal feasibility of the setup could be shown. The scanner is able to generate an FFL, to rotate it and to translate the line in any direction on the xy-plane. Here the scanner was powered with direct current and static measurements confirmed a good field quality. A gradient strength of 0.4 Tm^{-1} was used for testing.

In this work the expansion to a dynamic translation of the FFL is desired with a gradient strength of 1 Tm^{-1}. This value is the basis for all assumptions concerning power loss, current strength and coupling effects inside the scanner system.

The receive coils have been taken from the bachelor thesis [48]. As shown in Figure 3.7, a sinussoidal litz wire distribution was used. This geometry was proposed in [49] and grants a homogeneous sensitivity in the ROI, which is helpful in the reconstruction. As stated in 2.2.4, two receiving coil pairs are needed to acquire enough raw data for reconstruction. These two pairs lie orthogonal on a cylinder, where their symmetry axes meet at the main axis of the cylinder. The tube with the installed coils can be inserted into the gateway of the scanner.

3.2   Overview of the signal chain

This scanner was designed by the author prior this work. The implementation and a measurement of the generated fields was done in the master thesis [21]. With the provided coil assembly, the signal chain is the main topic of this work.

As already stated, a DF frequency $f_0 = 25$ kHz and a rotation frequency $f_{rot} = 100$ Hz are chosen. Following equation 3.3 this leads to 250 angles of data aquisition. Since this how-

Figure 3.6: The coil assembly is shown. The upper graphic displays the planned assembling and the lower picture is taken from the experimental setup. In the center of the build-up the coils are visible surrounded by the installed air cooling system.

ever is the first implementation of an FFL scanner with a rotatable line, the first experiments with the setup will be run at discrete angles of rotation. This is realized with direct current on the SF coils. That does increase the average power loss by a factor of $\sqrt{2}$ compared to alternating current in the SF coils, but provides the possibility of pauses between two angles of rotation.

Since the build-up of the scanner has been presented and the usage of direct current for first experiments was motivated, a short overview of the signal chain can be given and explained in detail in the subsequent sections.

The signal chain

A scanning progress is completely controlled by the user at a computer. Here the direct current for the SF coils can be set and the DF amplitude, frequency and phase can be controlled. After the SF is set and an FFL with a chosen rotation angle is generated, the DF translates the line over the FOV at the transmission frequency f_0.

This base frequency is generated by I/O cards. Here a signal with a perfect fundamental frequency $f_0 = 25$ kHz is assumed. To increase the signal to the desired voltage for the scanner, power amplifiers are necessary. Now, the signal is at a desired level, but the amplifiers pro-

Figure 3.7: Used receiving coils: Two orthogonal coil pairs on a tube with a discrete sinu-sodial litz wire distribution in the plane perpendicular to the main axis. The coils provide a very homogeneous field profile near the center.

duce higher harmonics of the fundamental frequency. To not detect these 'false signals', a filtering step is required. A bandpass filter was designed to damp the higher harmonics, while not influencing the fundamental frequency. The possibly clean signal then arrives at the DF coils and causes a particle response in the scanner. This signal can be recorded in the receiving coils.

Additionally to the particle response, the input signal induces a voltage in the receiving coils and in the SF coils. To prevent an alternating current in the SF coils, which would influence the static FFL, frequency stop filters were installed between the SF coils and their power sources. The damping of f_0 in the receiving chain is of great importance, since it dominates the detected particle responses by several orders of magnitude. Hence a bandstop filter is the first element in the receiving chain.

After this, a linear and a differential amplifier increment the signal to a fitting range for the I/O-card. Furthermore, the differential amplifier reduces coupling effects in the signal channels.

The then recorded raw data is saved and can be utilized for image reconstruction, given in section 2.2.4.

3.3 Transmission chain

To enable a completely computer controlled handling of the scanner, the fundamental frequency $f_0 = 25$ kHz is generated and the particle signal is recorded by an I/O card. The used card can produce voltages up to $V_{PP} = 10$ V. The signal is directly connected to an AE Techron 7796 power amplifier [50]. Here an amplification of 20V/V is provided.

The AE Techron 7796 has a gain linearity of 0.1%. That implicates that this power amplifier has at an input frequency f_0 not only an output signal with the frequency f_0, but also with higher odd harmonics of this fundamental frequency: $3f_0 = f_3, 5f_0 = f_5, 7f_0 = f_7$, etc. .

The problem here is that the later produced signal consists of these frequencies and the here

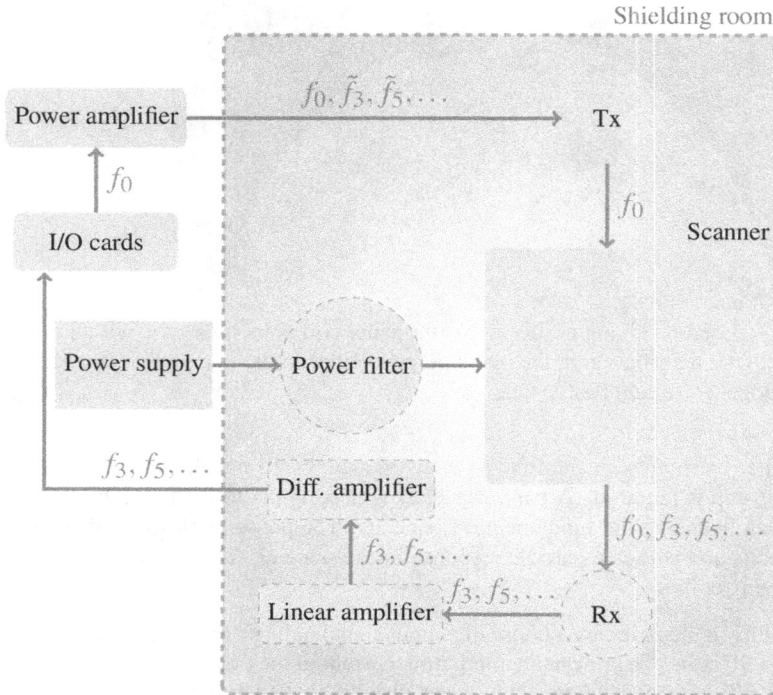

Figure 3.8: Overview of the MPI signal chain. At the top left the I/O-card generates the base frequency f_0, the power amplifier brings the amplitude to the desired level, but produces disturbing higher harmonics of the fundamental wave. These are damped by the Transmission filter (Tx). The particle signal, coming from the scanner, consists of the higher harmonics. f_0 also couples into the receiving coils and is filtered in the receiving filter (Rx). A linear amplifier brings the relatively small signal to the I/O-card level. The differential amplifier then creates a differential signal to avoid further coupling of disturbing signals. The I/O card then records the data. Further the power supply creates the SF and the power filter avoids coupling of alternating current from the DF into the power supply. Dashed lines symbolize shielding.

produced higher harmonics will be transmitted from the transmission coils into the receiving coils via induction. This means that a false signal is produced without any nanoparticles inside the scanner. Consequently, these frequencies have to be damped before the signal reaches the transmission coils.

The damping is performed by a 3rd order Butterworth bandpass filter [51]. This architecture is chosen over e.g. a Chebychev filter, since it provides a smooth characteristic without any oscillations in addition to the resonant case. This is profitable, since the damping of the higher harmonics has to be maximized and Butterworth filters allow an easy prediction of the damping of individual frequencies.

The design was achieved with the software Filter Free 2011. A center frequency of 25 kHz and a bandwidth of 8 kHz were used. This bandwidth provided an acceptable damping for the disturbing signals with realizable network components.

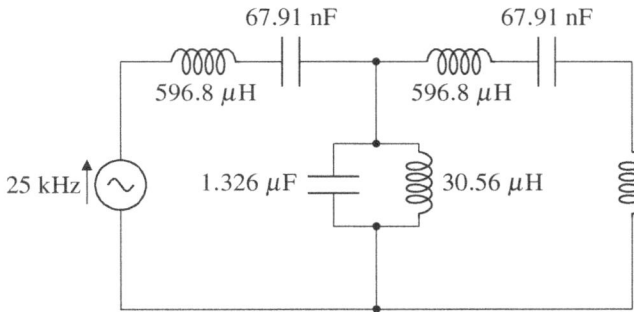

Figure 3.9: Circuit diagram of a 3rd order Butterworth bandpass filter. Each slope has a resonant frequency of $f_0 = 25$ kHz, leading to an overall resonant frequency of f_0. The serial resonant circuits act like short circuits at resonance and the parallel resonant circuit acts like a perfect open circuit. Hence signals with $f = f_0$ are transmitted without a loss in amplitude. All other signals experience damping.

The build-up of the filter is shown in Figure 3.9. Here the signal source is the AE Techron 7796 power amplifier, that provides the frequency f_0 (coming from the I/O-card) and the above discussed higher harmonics of the fundamental wave.

The first bench is a serial oscillating circuit, which is connected in series to the load (the transmission coil). The resonance frequency of this and each other individual bench lies at f_0. At this frequency, the first bench acts like a short circuit, for all other frequencies, damping is accomplished, as presented in section 2.3. Then a parallel LC-bench follows, parallel to the load. This dissipates all frequencies, except the resonance frequency f_0 to ground. After this another serial oscillation circuit, installed like the first bench follows to further damp the harmonics. This way, with every bench, the amplitudes of the harmonics decrease.

The simulated frequency response function of this filter shwon in figure 3.10 displays the damping of signals dependent on their frequencies.

The most important values in this graph are the damping for signals with frequencies at

- the 1. odd harmonic $f_3 = 75$ kHz (-55.227 dB) and

- the 2. odd harmonic $f_5 = 125$ kHz (-70.563 dB).

3.3.1 Impedance matching

Since the source resistance of the power amplifier and the load resistance of the DF coils are not the same, the power transmission is decreased [40].

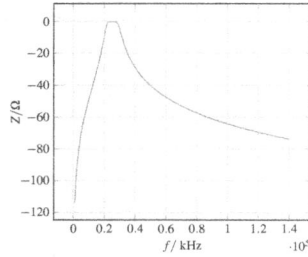

Figure 3.10: Simulated frequency transfer function for the transmission filters. The resonant case, where the damping is at its minimum, occurs at $f = 25$ kHz.

To overcome this problem, a so called L-network is added between load and filter. This can be used, when the load resistance is much smaller than the source resistance, which is the case in this network (see below). In this work the L-network consists of two capacities, one parallel to the load, one connected in series, displayed in Figure 3.11.

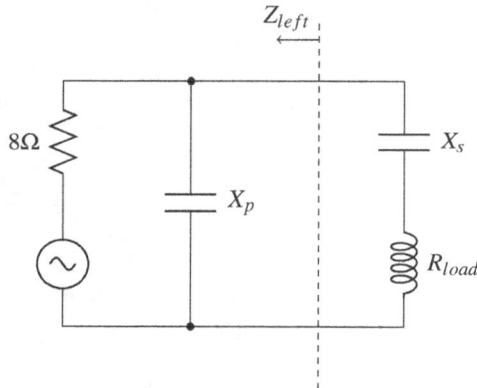

Figure 3.11: The circuit diagram of an L-network is displayed. It functions as an impedance matching between source and load and hence optimizes the power transfer in the network.

The goal of this network is to bring the load resistance to the same level as the source resistance. This is the resistance of the power amplifier and lies at $R_{left} = 8\Omega$. The load is the transmission coil inside the scanner. Due to their differing distance to the gateway, the DF coils for the x- and y-direction have slightly different shapes and hence slightly different properties

- $L^{outer} = 9.21\ \mu H$

- $R_s^{outer} = 20.9\ m\Omega$

- $L^{inner} = 10.99\ \mu H$

- $R_s^{inner} = 20.7\ m\Omega$

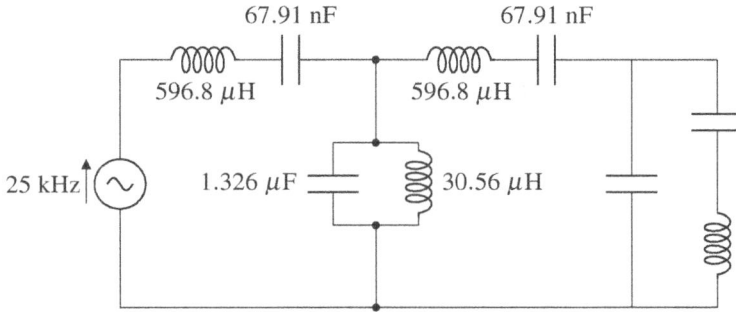

Figure 3.12: An impedance matching L-network has been added to the in Figure 3.9 given Butterworth filter. The two capacities allow a more efficient power transfer.

So for the two filters, the impedance matching had to be done individually.

In figure 3.11 an example network is shown. The impedance of the left-hand side is given by

$$Z_{left} = R_{left} + iX_{left} ,\qquad(3.4)$$

with i being the imaginary unit and X_{left} being the reactance (see section 2.3.3) of the left handed network [40]. The reactance of the parallel component X_p can be found by

$$R_{left} + iX_{left} = \frac{8\Omega i X_p}{8\Omega + iX_p} = \frac{(8\Omega)^2 i X_p + 8\Omega X_p^2}{(8\Omega)^2 + X_p^2} .\qquad(3.5)$$

The purpose of the last step was to create a real and an imaginary summand. This could be achieved by extending with $(8\Omega - iX_p)$. Now one can pick the value of X_p in a manner, that the real part of Z_{left} will be equal to R_{load}. With a capacitor C_p can then be found over the relation $X_p = \frac{1}{\omega C_p}$.

To get rid of the imaginary part in the result in equation 3.5 or the series reactance of the left handed network X_{left}, the series reactance X_s of the right handed network has to be chosen to let both add to zero

$$X_{left} = -X_s = \frac{1}{\omega C_s}\qquad(3.6)$$

This way, the capacities for the impedance match were found and the L-network was integrated into the transmission chain as displayed in the circuit diagram in Figure 3.12.

3.4   Receiving chain

The object of the receiving chain is to transmit the higher harmonics produced by the nanoparticles to the I/O card. The challenge is that the fundamental frequency f_0 inside the transmission coils induces a signal inside the receiving coils. This signal consists of the frequency f_0 and an amplitude, that can be calculated over the coupling factor between the coil assemblies.

The coupling factor has been measured by applying an alternating current to the DF coils and the signal amplitude has been measured at the receiving coils by an oscilloscope. The value is given in the results chapter 4.

Here, the in the receiving coil coupled signal with the fundamental frequency is several orders of magnitude higher than the generated particle signal. A 4th order Butterworth bandstop filter was chosen to block the input frequency f_0, while damping the higher odd harmonics to a minimum. The design was accomplished with the software Filter Design. The circuit diagram is displayed in Figure 3.13.

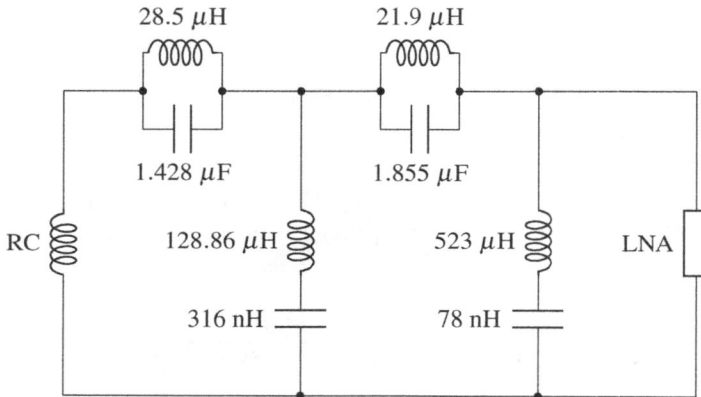

Figure 3.13: Butterworth bandstop filter: The whole filter and each individual slope has as resonant frequency of $f_0 = 25$ kHz. This way, the parallel resonant circuits (connected in series to the load) function as individual band stop filter and the serial resonant circuits (parallel to the load) ground signals with a frequency of f_0 and lead to an even stronger damping. The load is a linear amplifier (low noise amplifier, LNA), the source is a receiving coil (RC) in the scanner.

The principle for the damping mechanism is the same as in the bandpass filter. The main difference is the swap of parallel and serial networks. Now serial networks ground the fundamental frequency f_0 and parallel networks stop this frequency, connected in series to the load.

The first bench is thereby such a parallel resonant circuit, because a serial circuit parallel to the load would short-circuit the first coil.

The simulated influence of the filter on signals with specific frequencies can be seen in the

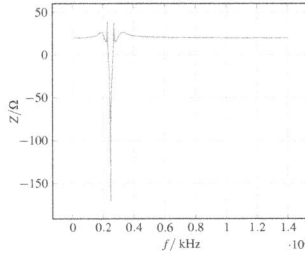

Figure 3.14: The simulated frequency response function for the constructed receiving filters is given. At the resonance frequency of $f = 25$ kHz, maximum damping is achieved. Thereby the frequencies, that serve for imaging are influenced to a minimum.

frequency response function in Figure 3.14. The realized values can be found in the results chapter 4.

3.5 Selection field signal

Beside the above presented main signal chain, the coupling of the transmission coils into each other has to be considered. Since for the first test, only direct current is applied to the SF coils, only the coupling of the DF frequency f_0 into the SF coils has to be considered.

To calculate the induced voltage, again the coupling factor between the two assemblies has been measured. This was done by applying an alternating current with frequency f_0 to the DF coils and the signal amplitude has been measured at the SF coils by an oscilloscope.

If no filtering would be performed, the induced voltage would produce an alternating current inside the SF coils. This current would not only be harmful for the power sources, but would also cause a decline of the SF quality. The second effect occurs, because the for a possibly good field quality adjusted currents inside the SF coils would be superimposed and the accuracy would get lost.

To stop the induced signal, a parallel resonant circuit was connected between the SF coils and their power sources. Since a high amplitude is expected, the coils could not be made of toroids, due to a too strong temperature development inside the toroids. A high surface to volume ratio was desired and could be achieved by freely winded coils.

The same coils, as in the above presented first bench of the receiving Butterworth filter were chosen. With the inductance given from these coils, the capacitance for the resonant circuit had to be chosen as

$$C = \left((2\pi f_0)^2 28.5 \ \mu\text{H}\right)^{-1} = 1.428 \ \mu\text{F}, \tag{3.7}$$

as given in the first bench of the 4th order Butterworth filter.

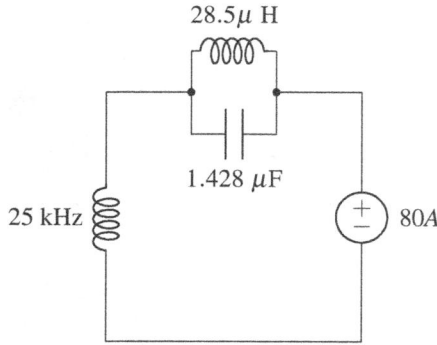

Figure 3.15: To prevent an alternating current inside the SF coils and their current source, a parallel resonant circuit is interconnected. At its resonant frequency 25 kHz, the current flow is damped to a maximum.

3.6   Practical realization

With the composed setup, the practical realization of the individual components was done. The I/O-cards were already installed and usable. In a first step, the DF power amplifiers (AE Techron 7796) were tested and integrated into the network.

The central part in the complete network are the filters. Their damping of the disturbing signals is critical for the noise level in the reconstruction of images.

To build these filters, the coils were produced in different ways. For a high inductance of a coil with a low signal level, toroidal inductors were used. To calculate their inductance, the approximation

$$L = N^2 \frac{\mu_0 \mu_r b}{2\pi} \ln \frac{R}{r} \tag{3.8}$$

can be used [52]. N is the number of windings and μ_0 and mu_r are the permeability in vacuum and in the used material. The parameters b, R, and r describe the thickness, the outer radius and the inner radius of the used toroid.

This procedure was not possible for coils with a too high signal level since the heat development inside the toroid would be too high. So freely winded coils were produced, using a self constructed negative, presented in Figure 3.16.

3.6.1   Shielding

In this work, only copper is used to shield electronic networks and hence variables presented in section 2.4.1 for calculating the skin effect and the accompanying shielding can be given.

The material constants of copper are:

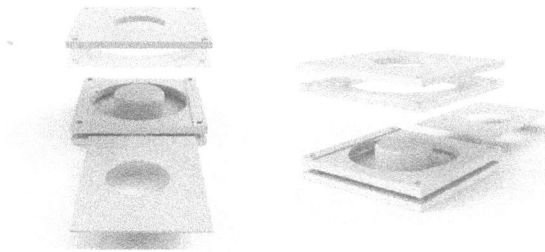

Figure 3.16: Coil form, that was used to produce inductors for the first order of the receiving filter and the SF power filter. Litz wire can be winded inside in two times two layers with a spacer in between. The spacer is build in to provide a higher surface to volume ratio.

- The permeability $\mu_r = 1 - 6.4 \cdot 10^{-6}\,\mathrm{N}\,\mathrm{A}^{-2}$

- The specific resistance $\rho = 1.678 \cdot 10^{-2}\,\Omega\,\mathrm{m}$

With this the depth of penetration of electromagnetic waves at frequencies, that could influence the signals in the network setup can be given. Since the penetration depth, that is calculated below, is inversely proportional to the angular frequency $\omega = 2\pi f$, it is sufficient to give the worst case of 25 kHz, which is the lowest frequency, that is to consider in terms of shielding. Coming from the considerations in equation 2.95, the skin depth, where the incoming electromagnetic wave is weakened by a ratio of e^{-1}, is

$$\delta = \sqrt{\frac{8.706\,\Omega\mathrm{m}^{-1} \cdot 0.001963\,\mathrm{mm}^2}{\pi \cdot 25\,\mathrm{kHz} \cdot 4\pi\,\mathrm{Hm}^{-1} \cdot 0.9999936\,\mathrm{Hm}^{-1}}} = 0.4161\,\mathrm{mm}\,. \tag{3.9}$$

The components, that require shielding are indicated in the signal chain overview in Figure 3.8. Copper plates and tubes with a wall thickness of 2 mm were used to guarantee a shielding for frequencies at and above 25 kHz.

The shielding concept of the receiving filter is shown in Figure 3.17. Here the first level of the Butterworth filter is placed inside a large tube, the three subsequent benches are located inside the smaller tubes. The following linear amplifier and the differential amplifier are placed inside the second large tube. This way, the signal receiving chain is shielded, starting at the Butterworth filter.

After this, the differential signal path makes shielding unnecessary, since coupling disturbing signals couple into both paths of the differential conductor and cancel each other out.

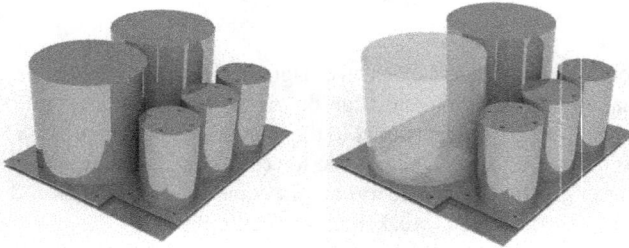

Figure 3.17: Shielding of individual filter levels for the Butterworth bandstop filter. The first resonant circuit is located in a large tube. Next the signal passes the three small tubes, where the next three levels of filtering are performed. After this the signal is transmitted to a low noise amplifier and next to a power amplifier in the second large tube. Here the amplitude is set to I/O-card level.

Chapter 4

Results

At the end of this work, the whole setup, as shown in figure 4.1 is implemented. The scanner features a discrete rotation of the FFL and a dynamic translation of this line using the DF coils.

The individual results are structured in the same order, as the signals pass the signal chain. Starting at the I/O cards, the signal with a frequency $f_0 = 25$ kHz is transmitted to the power amplifier. Here the signal amplification of a factor of 20V/V could be confirmed.

The component as part of this thesis is the transmission filter, which will be presented in the upcoming section.

4.1   Transmission filter

The 3rd order Butterworth bandpass filter presented in section 3.3 with a center frequency of $f_0 = 25$ kHz and a bandwidth of 8 kHz could successfully be implemented.

The physical composition of one of the two filter is demonstrated in Figure 4.2. Here the toroidical inductors are visible, assembled to vertical plates. The distance between the coils was chosen to be higher than the outer radius of the coils. This distance served as a guidance value to prevent coupling in between the coils. Next to the coils and mounted onto circuit boards are the capacitors.

To provide a solid grounding, all conductors at mass level are connected at one central node.

The function of the filter is to block all disturbing signals, namely the higher harmonics, generated by the power amplifier described in section 3.3. The simulated frequency transfer function together with the simulated phase shift as a function of the frequency are shown in Figure 4.3.

After their implementation, the transmission filters for x- and y-direction were examined with a network analyzer. The resulting frequency transfer functions with the corresponding phase shift are shown in Figure 4.4.

Figure 4.1: The complete setup. The scanner is displayed, next to a robot, that can position probes inside the scanner. Directly under the scanner, the power filters are visible. To the left of these filters follow the transmission filter in white and the receiving filter, shielded with copper.

In the simulation, a damping of the first two disturbing frequencies f_3 and f_5 were predicted:

Damping $d(f)$ of a signal

- at f_3: $d(f_3) = d(75 \text{ kHz}) = -55.227$ dB

- at f_5: $d(f_5) = d(125 \text{ kHz}) = -70.563$ dB

The actually realized damping in the filter for x- and y-direction in comparison to these simulated values are listed below. Due to the method of measuring, closely neighboring values are used as upper and lower bounds.

The constrains are:

- -49.877 dB $< d_x(f_3) < -50.325$ dB

- -64.747 dB $< d_x(f_5) < -65.018$ dB

- -49.660 dB $< d_y(f_3) < -50.106$ dB

- -64.363 dB $< d_y(f_5) < -64.646$ dB

Taking the worst case, absolute errors of between 5.35 and 6.2 dB occur for each filter and analyzed frequency, leading to relative errors between 8.2% and 10%. These errors were predictable, since the filter design and simulation programs had no information about the resistances of the individual components, as described in the discussion.

Figure 4.2: The physical implementation of the 3rd order Butterworth bandpass filter for the transmission signal is shown. The toroid coils are assembled to vertical plates and the yellow and blue capacitors are mounted onto circuit boards.

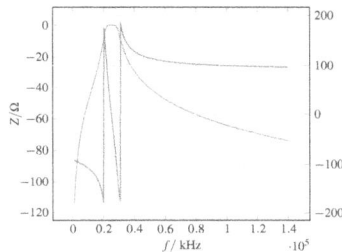

Figure 4.3: The simulated frequency transfer function is illustrated in red and the frequency-dependent phase shift is displayed in blue. The resonant frequency lies at $f_0 = 25$ kHz and the bandwidth of 8 kHz is also visible.

The accomplished damping of about 50 dB for signals with a frequency of f_3 means that the amplitude of these disturbing signals is lowered by a factor of 10^{-5}.

To calculate the signal amplitude of the disturbing signals, the voltage input for the amplifiers is needed. For simplification the listing of all currents and voltages in all components is left out. Instead a worst case scenario is described by taking an input voltage of 4 V_{pp} at the amplifiers. With the amplification factor of 20 V/V the output voltage for the fundamental frequency is 80 V_{pp}. Featuring a cumulative noise level at 8Ω of 0.77% of the output amplitude the amplifiers emit noise signals, including the discussed higher harmonics, with a amplitude sum of 632 mV_{pp}.

Since this is a sum of amplitudes the expected level for a single disturbing frequency is expected to be lower than this. The calculated value however serves as a worst case value for the first higher odd harmonic of the fundamental frequency.

Applying the above given damping of the implemented transmission filters this signal amplitude is lowered to 6.32 μV. The voltage induced by this signal inside the receiving coils is

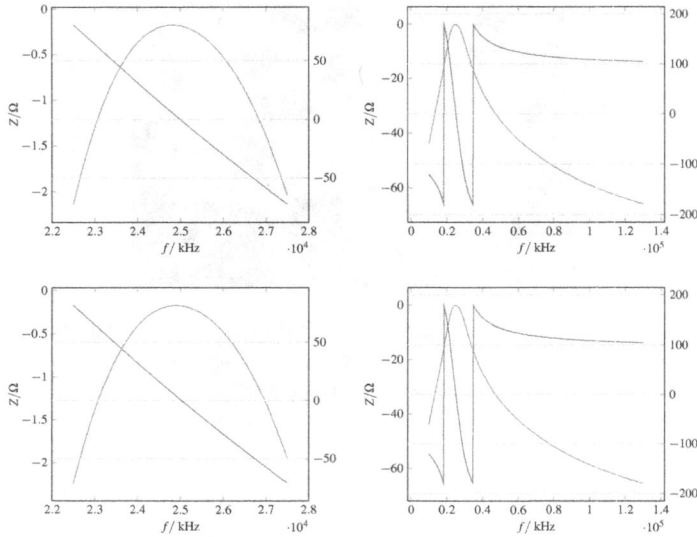

Figure 4.4: The measured transfer functions of the 3rd order Butterworth bandpass filter in x- and y-direction are displayed in the upper and lower row. Each red graph gives the frequency transfer functions and the blue graphs show the corresponding phase shift.

$$\tilde{U}_{\text{induced}} = 6.32\ \mu\text{V} \cdot \frac{1\ \text{V}}{40\ \text{V}} = 158\ \text{nV}\,. \tag{4.1}$$

Since for this worst case scenario the sum of all disturbing signal amplitudes was used for the first odd harmonic, the actually appearing noise level is expected to be much lower.

4.2   Receiving filter

The more complex build-up of the receiving filter could also be accomplished. As shown in Figure 4.5, the setup could be realized, as planed in section 3.6.

The analysis of the filter has been done analogue to the analysis of the transmission filter. A frequency transfer function and a corresponding phase shift curve have been recorded and compared to the simulated curves. The simulation is shown in Figure 4.6 and the actually measured data is displayed in Figure 4.7.

The simulation predicts a damping of -170.72 dB at the resonant frequency of 25 kHz.

The damping achieved in x- and y- direction are -79.16 dB and -81.15 dB. This covers only about 50% of the desired damping. Nevertheless the damping in this order of magnitude is sufficient, when considering induced signal amplitude predicted over the coupling factor.

Figure 4.5: The physical realization of the 4th order Butterworth bandstop filter is shown. The large tubes accommodate the first level of the filter and the linear and differential amplifiers. In the small tubes, the second to forth order of filtering takes place.

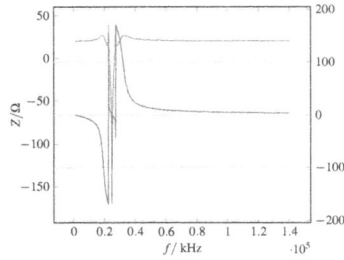

Figure 4.6: Simulated frequency transfer function and phase shift behavior of the bandstop filter for the receiving coils in x- and y-direction. The object is to damp the center frequency f_0, while providing a constant progress of the curve for the particle signals.

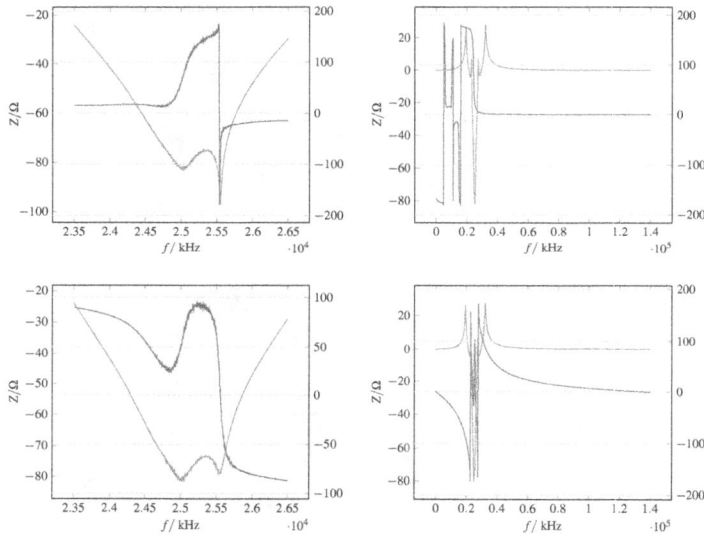

Figure 4.7: With a network analyzer measured frequency transfer function of the implemented receiving filter. The propose of the filter is to damp the coupling signal with frequency f_0, while influencing signals at other frequencies to a minimum. The upper curves show the filter behavior for the x-direction, the lower curves for the y-direction.

The voltage applied at the DF coils in all cases lies 180 V. The measured coupling factor between the DF coils and the receiving coils lies at the worst case at $\frac{1\,V}{40\,V}$. This means, that the expected amplitude of a signal with frequency f_0 in the receiving channel is always below 4.5 V. A damping by 80 dB means a reduction of this amplitude to $4.5 \cdot 10^{-8}$ V = 45 nV.

What is noticeable in the measurements in Figure 4.7 are the side oscillations of the phase shift for the filter in x-direction at frequencies below 25 kHz. In addition to that in the left column of Figure 4.7 shows an oscillations at about 25.5 kHz. As discussed in chapter 5, these effects are measurement errors.

Hence, the goal of damping the coupled fundamental wave was accomplished. Also successful was the minimal influence on the higher harmonics of this frequency. The frequency transfer function has a nearly constant progress towards higher frequencies, staying between 0.22 dB and 0 dB in the analyzed cases.

4.3 Power filter

The protection of the SF circuit against the coupling of the DF signal was achieved by the implementation of the shielded power filters. Since four independent circuits exist, each one supplying a Maxwell coil pair with direct current, also four filter had to be realized.

As displayed in Figure 4.8, the freely winded coils were successfully produced and the resonant case could be reached by applying capacitors parallel to the coils. These resonant circuits were build inside a copper shielding. This way the disturbing signal could be canceled.

Figure 4.8: The parallel resonant circuits, damping the coupling of the DF signal into the SF coils, could be realized and shielded. Four filters were necessary, one for each Maxwell coil pair, generating and rotating the FFL.

Chapter 5

Discussion

The signal chain could be completely implemented in this work. However no reconstructed images are presented yet. This occurs due to a problem in the not in this work implemented but used linear amplifiers. In this chapter, this problem is discussed, issues concerning individual parts of the signal chain are argued and an outlook of future work is given.

The transmission chain was successfully planed and realized. It is able to provide only signals with the desired frequency and damps all disturbing signals, that could be produced by the power amplifiers or by coupling effects, by at least 5 orders of magnitude.

In the simulation, a damping of 5.5 orders of magnitude was calculated. This value was by slightly undermatched, because the filter designing and simulating software had no information about the resistance of the individual network components. Consequently a lowering of the quality of the resonant circuits was expected. Still a damping of about 90% of the desired factor was reached.

The sufficiency of 50 dB damping of disturbing signals has to be discussed. As calculated in equation 4.1 in the results chapter, the sum of all disturbing signals and of all noise, emitted from the amplifier that actually reaches the receiving chain features an amplitude of 158 nV. This means that the amplitudes of the individual signals will be lower than this upper bound.

For a particle signal good enough to enable reconstruction, its amplitude has to feature a similar or higher level. Equation 2.60 gives the computation of the particle signal inside the receiving coils. This equation can for the presented case be simplified. A homogeneous receiving coil sensitivity $\mathbf{p} = p$ given through the utilized coil shape can be assumed. Further a homogeneous external magnetic field and a constant concentration distribution of the probe inside its volume V_p are assumed. This simplifies equation 2.60 to

$$u(t) = -\mu_0 p V_p c \frac{\partial}{\partial t} \overline{\mathbf{m}}(\mathbf{H}_{DF}(t)).$$ (5.1)

This equation shows that the particle signal can easily be increased by taking slightly larger diameters D of tracer phantoms. The induced voltage scales with the 3rd order of D and can this way be set larger than the noise level. This is of course critical for a clinical approach, but sufficient enough for testing the feasibility of the system. Hence, a phantom with differ-

59

ent tracer volumes is in production. The future work of this scanner, presented at the end of this chapter will this way be realizable.

With the above discussed particle responses, the damping of the fundamental frequency f_0 inside the receiving chain can also be analyzed. As presented in the results, f_0 is damped by 8 orders of magnitude and hence has an amplitude of 45 nV_{pp} in the worst case. Compared to the scalable particle signal, this can be considered as a sufficient damping.

Also the behavior of the filters for the particle signal has been implemented as desired. The constant progress of the frequency transfer function and its proximity to 0 dB are optimal conditions for image reconstruction.

The second resonant cases at 25.5 kHz can be neglected, since no signal with this frequency is to expect in the signal chain. Also the additional zero-crossing of the phase shift in the transfer function of receiving filter for x-direction (see Figure 4.7 at the top right) can be neglected for the same reasoning. In addition several followup measurements have been made and the presented resonant cases were either missing or displaced to lower frequencies. This could be traced back to measurement errors. The contacts of the used network analyzers proved to be unreliable. For the in the results chapter presented values of the frequency transfer function, the followup measurements showed constant results. Hence, the measured findings are trustworthy.

With the so far discussed components, no problems occur for image reconstruction. The next component, following the receiving filter is a linear amplifier. The object of this component is to increase the amplitude of the relatively small particle signals to a level, where they can be recorded by the I/O-card. Here a coupling problem occurs.

A signal with a frequency $\tilde{f} = 4$ MHz appears behind the amplifier. The source of this signal is either a coupling effect, with an external source producing the signal. As rigorous shielding experiments show, the signal is transmitted by the electric circuit, which supplies the power outlets and lamps inside the shielding cabin. To generate a measurable particle signal inside the receiving chain, this problem has to be overcome first.

A first approach will be to run the system without any power supply coming from inside the cabin. Since the transmission amplifiers, I/O-cards, and SF power supply are located outside the cabin, only the power supply for the amplifiers at the receiving chain have to be replaced by batteries. The second approach will be to purify the electric circuit inside the cabin.

As soon as this problem is overcome, experiments with the system are prospective.

As a future work, the reconstruction of phantoms will be the first step. After this, different reconstruction algorithms considering the spatial encoding technique of the field free line can be programmed and evaluated.

Furthermore, the coil assembly is feasible of generating a field free point and also to move this point in a programmable trajectory over a probe. This means that a data acquisition with the FFP as a spatial encoding scheme is possible. The generated raw data can be used for reconstruction and a direct comparison between the FFP and the FFL encoding scheme will be possible.

After this, the scanner might be reworked to enable a dynamic selection field. This means that issues concerning the coupling of the selection field signals inside the receiving and drive field coils have to be concerned. However this also means that the average power loss inside the selection field coils will decrease by a factor of $\sqrt{2}$ and hence a higher gradient is prospective.

Chapter 6

Conclusion

An MPI system with a dynamic field free line was realized. The main results can be divided into the implementation of the transmission chain and the receiving chain.

On the transmission side, the signal could successfully be amplified and the disturbing signals generated during the amplification could be damped to a desired level. To achieve this, a 3rd order Butterworth bandpass filter was designed and implemented. With the existing coil assembly the signal generating part of the system is complete.

The receiving coils were integrated into the scanner and connected to the receiving chain. Here, signals with the drive field frequency could be damped as well, while influencing the particle signals to a minimum. Therefor a 4th order Butterworth bandstop filter was designed, realized and shielded against electromagnetic induction.

To stop a coupling of the drive field signals into the selection field coils, several parallel resonant circuits have been built and connected to the selection field generating circuit.

However it was not possible to reconstruct images yet. Since the particle signal has a low amplitude, an amplification component follows in the signal chain. After this linear amplifier, a disturbing signal with a frequency of 4 MHz is amplified in the signal chain. The source of this signal could be identified as the electric circuits for the lamps and power outlets of the shielding cabin. Until this signal is eliminated, the system will be run without using the affected circuit, which will be realized in a future work.

As future experiments with this scanner, image generation using different reconstruction schemes and a direct comparison between the FFL and FFP spatial encoding scheme are prospective.

Chapter 7

Bibliography

[1] B. Gleich and J. Weizenecker. Tomographic imaging using the nonlinear response of magnetic particles. *Nature*, 435:1214–1217, 2005.

[2] J. BORGERT, B. GLEICH, and T. M. BUZUG. *Handbook of Medical Technology. 1st Edition.* Berlin: Springer-Verlag, 2011, Chap. Magnetic Particle Imaging, pp. 461–476.

[3] D. FINAS, K. BAUMANN, K. HEINRICH, B. RUHLAND, L. SYDOW, T. F. SAT-TEL, K. GRÄFE, K. LÜDTKE-BUZUG, and T. M. BUZUG. Distribution of super-paramagnetic nanoparticles in lymphatic tissue for sentinel lymph node detection in breast cancer by magnetic particle imaging. *SPPHY*, 140:187–191, 2012.

[4] K. GRÄFE, T. F. SATTEL, K. LÜDTKE-BUZUG, D. FINAS, J. BORGERT, and T. M. BUZUG. Magnetic particle imaging for sentinel lymph node biopsy in breast cancer. *SPPHY*, 140:237–241, 2012.

[5] J. HAEGELE, J. RAHMER, B. GLEICH, C. BONTUS abd J. BORGERT, H. WO-JTCZYK, T. M. BUZUG, J. BARKHAUSEN, and F. M. VOGT. Visualization of instruments for cardiovascular intervention using mpi. *SPPHY*, 140:211–215, 2012.

[6] J. WEIZENECKER, B. GLEICH, J. RAHMER, H. DAHNKE, and J. BORGERT. Three-dimensional real-time in vivo magnetic particle imaging. *Physics in Medicine and Biology*, 54(5), 2009.

[7] T. M. BUZUG, T. F. SATTEL, M. ERBE, S. BIEDERER, D. FINAS, K. DIEDRICH, F. VOGT, J. BARKHAUSEN, J. BORGERT, K. LÜDTKE-BUZUG, and T. KNOPP. *Nanomedicine – Basic and Clinical Applications in Diagnostics and Therapy.* Karger, 2011.

[8] T. BUZUG, J. BORGERT, T. KNOPP, S. BIEDERER, T. SATTEL ad M. ERBE, and K. LÜDTKE-BUZUG (EDS.). *Magnetic Nanoparticles: Particle Science, Imaging Technology, and Clinical Applications.* World Scientific Publishing Company, 2010.

[9] R. M. Ferguson, K. R. Minard, A. P. Khandhar, and K. M. Krishnan. Optimizing magnetite nanoparticles for mass sensitivity in magnetic particle imaging. *Medical Physics*, 38(3):1619– 1626, 2011.

[10] S. BIEDERER, F. M. VOGT, K. LÜDTKE-BUZUG, T. KNOPP, T. F. SATTEL, J. BARKHAUSEN, and T. M. BUZUG. A study on the performance of different super-paramagnetic iron oxide particles in magnetic particle imaging. *Magnetic Resonance Materials in Physics, Biology and Medicine*, 1, 2009.

[11] J. HAEGELE, F. M. VOGT, J. BARKHAUSEN, T. M. BUZUG, and K. LUEDTKE-BUZUG. Eisenoxidnanopartikel für magnetic particle imaging (mpi). *Deutscher Röntgenkongress*, 93, 2012.

[12] P. W. GOODWILL, A. TAMRAZIAN, L. R. CROFT, C. D. LU, E. M. JOHNSON, R. PIDAPARTHI, R. M. FERGUSON, A. P. KHANDHAR, K. M. KRISHNAN, and S. M. CONOLLY. Ferrohydrodynamic relaxometry for magnetic particle imaging. *Applied Physics Letters*, 98(26):262502 –262502–3, 2011.

[13] S. BIEDERER, T. F. SATTEL, T. KNOPP, K. LÜDTKE-BUZUG, B. GLEICH, J. WEIZENECKER, J. BORGERT, and T. M. BUZUG. The influence of the particle-size distribution on the image resolution in magnetic particle imaging. *Magnetic Resonance Materials in Physics, Biology and Medicine*, 1, 2009.

[14] J. Weizenecker, B. Gleich, and J. Borgert. Magnetic particle imaging using a field free line. *Journal Of Physics D: Applied Physics*, 41(10):105009, 2008.

[15] T. Knopp, T. F. Sattel, S. Biederer, and T. M. Buzug. Field-free line formation in a magnetic field. *Journal Of Physics A: Mathematical And Theoretical*, 43:012002, 2010.

[16] T. Knopp, M. Erbe, S. Biederer, T. F. Sattel, and T. M. Buzug. Efficient generation of a magnetic field-free line. *Medical Physics*, 37(7):3538–3540, 2010.

[17] T. Knopp, M. Erbe, T. F. Sattel, S. Biederer, and T. M. Buzug. A fourier slice theorem for magnetic particle imaging using a field-free line. *Inverse Problems*, 27(9):095004, 2011.

[18] T. M. BUZUG, G. BRINGOUT, M. ERBE, K. GRÄFE, M. GRAESER, M. GRÜTTNER, A. HALKOLA, T. F. SATTEL, W. W. TENNER, J. HÄGELE, F. M. VOGT, J. BARKHAUSEN, and K. LÜDTKE-BUZUG. Magnetic particle imaging: Introduction to imaging and hardware realization. *Medical Physics*, 22(4):323–334, 2012.

[19] P. W. Goodwill and S. M. Conolly. Multidimensional x-space magnetic particle imaging. *IEEE Transactions on Medical Imaging*, 30(9):1581–1590, 2011.

[20] M. Erbe, T. Knopp, T. F. Sattel, S. Biederer, and T. M. Buzug. Experimental generation of an arbitrarily rotated field-free line for the use in magnetic particle imaging. *Medical Physics*, 38:5200, 2011.

[21] M. Weber. Power loss optimized field free line generation for magnetic particle imaging. Master's thesis, University of Lübeck, 2012.

[22] W. Nolting. *Basic Course Theoretical Physics 3: Electrodynamics; Grundkurs Theoretische Physik 3: Elektrodynamik*. Springer, Berlin, 2001.

[23] R. Lawaczeck, H. Bauer, T. Frenzel, M. Hasegawa, Y. Ito, K. Kito, N. Miwa, H. Tsutsui, H. Vogler, and H. J. Weinmann. Magnetic iron oxide particles coated with carboxydextran for parenteral administration and liver contrasting. *Pre-clinical profile of SH U555A. Acta Radiol*, 38(4 Pt 1):584–597, 1997.

[24] J. Weizenecker, J. Borgert, and B. Gleich. A simulation study on the resolution and sensitivity of magnetic particle imaging. *Physics In Medicine And Biology*, 52(21):6363, 2007.

[25] H. Landolt and R. Börnstein. *Numerical Data and Functional Relationsships in Science and Technology (New series III/4b Magnetic Oxides and Related Compounds)*. Springer, Berlin, 1977.

[26] P. W. GOODWILL and S. M. CONOLLY. The x-space formulation of the magnetic particle imaging process: 1-d signal, resolution, bandwidth, snr, sar, and magnetostimulation. *IEEE Transactions on Medical Imaging*, 29(11):1851–1859, 2010.

[27] J. J. KONKLE, P. W. GOODWILL, O. M. CARRASCO-ZEVALLOS, and S. M. CONOLLY. Projection reconstruction magnetic particle imaging. *IEEE Transactions on Medical Imaging*, 32(2):338–347, 2013.

[28] T. KNOPP, T. F. SATTEL, and T. M. BUZUG. Efficient positioning of the field- free point in magnetic particle imaging. *SPPHY*, 140:161–165, 2012.

[29] T. M. Buzug. *Computed Tomography: From Photon Statistics to Modern Cone-beam CT*. Springer-Verlag, Berlin Heidelberg, 2008.

[30] J. Radon. About the determination of functions by integral values along certain manifolds; über die bestimmung von funktionen durch ihre integralwerte längs gewisser mannigfaltikeiten. *Berichte über die Verhandlungen der Königlich-Sächsischen Gesellschaft der Wissenschaften zu Leipzig*, 69:262–277, 1917.

[31] P. W. GOODWILL, J. J. KONKLE, B. ZHENG, E. U. SARITAS, and S. T. CONOLLY. Projection x-space magnetic particle imaging. *IEEE Transactions on Medical Imaging*, 32(5):1076–1085, 2012.

[32] J. Rahmer, J. Weizenecker, B. Gleich, and J. Borgert. Signal encoding in magnetic particle imaging: Properties of the system function. *BMC Medical Imaging*, 9(4), 2009.

[33] M. GRÜTTNER, M. GRAESER, S. BIEDERER, T. SATTEL, H. WOJTCZYK, W. TENNER, T. KNOPP, B. GLEICH, J. BORGERT, and T. M. BUZUG. 1d-image reconstruction for magnetic particle imaging using a hybrid system function. *Proc. IEEE Nuc. Sci. Symp. Med. Im. Conf*, 2011.

[34] T. KNOPP, S. BIEDERER, T. SATTEL, J. RAHMER, J. WEIZENECKER, B. GLEICH, J. BORGERT, and T. M. BUZUG. 2d model-based reconstruction for magnetic particle imaging. *Medical Physics*, 37:485–491, 2010.

[35] M. GRAESER, T. F. SATTEL, M. GRÜTTNER, H. WOJTCZYK, G. BRINGOUT, W. TENNER, and T. M. BUZUG. Determination of system functions in magnetic particle imaging. *SPPHY*, 140:59–64, 2012.

[36] T. KNOPP, T. SATTEL, S. BIEDERER, J. RAHMER, J. WEIZENECKER, B. GLEICH, J. BORGERT, and T. BUZUG. Model-based reconstruction for magnetic particle imaging. *IEEE Transactions on Medical Imaging*, 29(1):12–18, 2010.

[37] T. KNOPP, S. BIEDERER, T. SATTEL, M. ERBE, and T. M. BUZUG. Prediction of the spatial resolution of magnetic particle imaging using the modulation transfer function of the imaging process. *IEEE Transactions on Medical Imaging*, 30(6):1284–1292, 2011.

[38] T. KNOPP, S. BIEDERER, T. SATTEL, and T. M. BUZUG. Singular value analysis for magnetic particle imaging. *Proc. IEEE Nuc. Sci. Symp. Med. Im. Conf.*, 2008.

[39] M. GRÜTTNER, T. F. SATTEL, F. GRIESE, and T. M. BUZUG. System matrices for field of view patches in magnetic particle imaging. *SPIE Medical Imaging*, 2013.

[40] J. B. Hagen. *Radio-Frequency Electronics*. Cambridge University Press, Cambridge, 2009.

[41] J. C. Maxwell. A treatise on electricity and magnetism. *Oxford: Clarendon*, 1873.

[42] T. M. BUZUG, T. F. SATTEL, M. ERBE, S. BIEDERER, M. GRAESER, M. GRÜTTNER, W. TENNER, H. WOJTCZYK, J. BORGERT, D. FINAS, K. DIETRICH, F. M. VOGT, J. BARKHAUSEN, K. LÜDTKE-BUZUG, and T. KNOPP. Magnetic particle imaging: Novel field generating devices for optimized imaging. *Biomed Tech*, 56, 2011.

[43] M. Erbe, T. F. Sattel, T. Knopp, S. Biederer, and T. M. Buzug. An optimized field free line scanning device for magnetic particle imaging. *Biomed Tech*, 56(1):298, 2011.

[44] T. KNOPP, T. F. SATTEL, and T. M. BUZUG. Efficient magnetic gradient field generation with arbitrary axial displacement for magnetic particle imaging. *Magnetics Letters, IEEE*, 3:6500104, 2012.

[45] T. F. SATTEL, M. ERBE, and T. M. BUZUG. Optimization of circular current distributions for magnetic field generation in mpi: A comparison of the selection field coil and the drive field coil geometry. *SPPHY*, 140:313–318, 2012.

[46] M. ERBE, M. WEBER, T. F. SATTEL, and T. M. BUZUG. Experimental validation of an assembly of optimized curved rectangular coils for the use in dynamic field free line magnetic particle imaging. *Current Medical Imaging Reviews*, 9(2):89–95(7), 2013.

[47] D. M. Ginsberg and M. J. Melchner. Optimum geometry of saddle shaped coils for generating a uniform magnetic field. *Review of Scientific Instruments*, 41(1):122–123, 1970.

[48] S. E. Heinitz. Realization of a tube-shaped receiving coil set for magnetic particle imaging with homogeneous sensitivity profiles; realisierung eines röhrenartigen empfangsspulensets für magnetic particle imaging mit homogenen sensitivitätsprofilen. Master's thesis, University of Lübeck, 2012.

[49] T. F. Sattel. Set of tube-like coils with orthogonal sensitivity profile for 3d magnetic particle imaging. *Manuscript, University of Luebeck*, 2011.

[50] L. Shank. *AE Techron 7548/7796 Operator's Manual*. AE Techron, AE Techron, Inc.; 2507 Warren Street, Elkhart, IN 46516, 04 2011.

[51] L. von Wangenheim. *Analogue Signal Processing: System Theory, Electronics, Filters, Oscillators, Simulation Techniques; Analoge Signalverarbeitung: Systemtheorie, Elektronik, Filter, Oszillatoren, Simulationstechnik*. Vieweg+Teubner, Wiesbaden, 2012.

[52] K. Küpfmüller. *Introduction to Theoretical Electro Engeneering; Einführung in die theoretische Elektrotechnik*. Springer, Berlin, 1990.

Infinite Science Publishing provides a publication platform
for excellent theses as well as scientific monographies and
conference proceedings for reasonable costs.

These publications enable scientists and research organizations
to reach the maximum attention for their results.

The service of Infinite Science Publishing comprises the entire
range from the publication of print-ready documents up to
cover design as well as copy-editing of single articles.

Infinite Science Publishing is an imprint of the Infinite Science
GmbH, a University of Lübeck spin-off and service partner of the
BioMedTec Science Campus.

www.infinite-science.de/publishing

Infinite Science GmbH
MFC 1 | BioMedTec Wissenschaftscampus
Maria-Goeppert-Str. 1, 23562 Lübeck
book@infinite-science.de

Infinite Science
Publishing

More Titles on Magnetic Particle Imaging

Herstellung und Charakterisierung superparamagnetischer Lacke
Inga Christine Kuschnerus
EUR 29,90

Optimierung der Permanentmagnetengeometrie zur Generierung eines Selektionsfeldes für Magnetic Particle Imaging
Matthias Weber
EUR 29,90

Compressed Sensing und Sparse Rekonstruktion bei Magnetic Particle Imaging
Anselm von Gladiß
EUR 49,90

Beschleunigtes Magnetic Particle Imaging: Untersuchung des Einsatzes von Compressed Sensing für die Signalaufnahme
Nadine Traulsen
EUR 49,90

Power-Loss Optimized Field-Free Line Generation for Magnetic Particle Imaging
Matthias Weber
EUR 49,90

Elliptical Coils in Magnetic Particle Imaging
Christian Kaethner
EUR 49,90

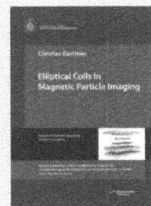

www.ingramcontent.com/pod-product-compliance
Lightning Source LLC
Chambersburg PA
CBHW081110220326
41598CB00038B/7304